NEUROSCIENCE RESEARCH PROGRESS SERIES

THE IMPORTANCE OF COLOUR PERCEPTION TO ANIMALS AND MAN

NEUROSCIENCE RESEARCH PROGRESS SERIES

Current Advances in Sleep Biology
Marcos G. Frank (Editor)
2009. ISBN: 978-1-60741-508-4

Dendritic Spines: Biochemistry, Modeling and Properties
Louis R. Baylog (Editor)
2009. ISBN: 978-1-60741-460-5

Relationship between Automatic and Controlled Processes of Attention and Leading to Complex Thinking
Rosa Angela Fabio
2009. ISBN: 978-1-60741-810-8

Neurochemistry: Molecular Aspects, Cellular Aspects and Clinical Applications
Anacleto Paços and Silvio Nogueira (Editors)
2009. ISBN: 978-1-60741-831-3

Sleep Deprivation: Causes, Effects and Treatment
Pedr Fulke and Sior Vaughan (Editors)
2009. ISBN: 978-1-60741-974-7

Color Perception: Physiology, Processes and Analysis
Darius Skusevich and Petras Matikas (Editors)
2009. ISBN: 978-1-60876-077-0

The Importance of Colour Perception to Animals and Man
Birgitta Dresp-Langley and Keith Langley
2010. ISBN: 978-1-60876-394-8

NEUROSCIENCE RESEARCH PROGRESS SERIES

THE IMPORTANCE OF COLOUR PERCEPTION TO ANIMALS AND MAN

BIRGITTA DRESP-LANGLEY
AND
KEITH LANGLEY

Nova Science Publishers, Inc.
New York

Copyright © 2010 by Nova Science Publishers, Inc.

All rights reserved. No part of this book may be reproduced, stored in a retrieval system or transmitted in any form or by any means: electronic, electrostatic, magnetic, tape, mechanical photocopying, recording or otherwise without the written permission of the Publisher.

For permission to use material from this book please contact us:
Telephone 631-231-7269; Fax 631-231-8175
Web Site: http://www.novapublishers.com

NOTICE TO THE READER

The Publisher has taken reasonable care in the preparation of this book, but makes no expressed or implied warranty of any kind and assumes no responsibility for any errors or omissions. No liability is assumed for incidental or consequential damages in connection with or arising out of information contained in this book. The Publisher shall not be liable for any special, consequential, or exemplary damages resulting, in whole or in part, from the readers' use of, or reliance upon, this material.

Independent verification should be sought for any data, advice or recommendations contained in this book. In addition, no responsibility is assumed by the publisher for any injury and/or damage to persons or property arising from any methods, products, instructions, ideas or otherwise contained in this publication.

This publication is designed to provide accurate and authoritative information with regard to the subject matter covered herein. It is sold with the clear understanding that the Publisher is not engaged in rendering legal or any other professional services. If legal or any other expert assistance is required, the services of a competent person should be sought. FROM A DECLARATION OF PARTICIPANTS JOINTLY ADOPTED BY A COMMITTEE OF THE AMERICAN BAR ASSOCIATION AND A COMMITTEE OF PUBLISHERS.

LIBRARY OF CONGRESS CATALOGING-IN-PUBLICATION DATA
Dresp-Langley, Birgitta.
 The biological significance of colour perception / Birgitta Dresp-Langley, Keith Langley.
   p. ; cm.
 Includes bibliographical references.
 ISBN 978-1-60876-394-8 (softcover)
 1. Color vision. 2. Physiology, Comparative. I. Langley, Keith. II. Title.
 [DNLM: 1. Color Perception--physiology. 2. Physiology, Comparative. WW 150 D773b 2009]
 QP483.D74 2009
 612.8'4--dc22
                    2009036574

Published by Nova Science Publishers, Inc. ✥ *New York*

CONTENTS

Preface		vii
Authors' Biographical Details		ix
Summary		xi
Chapter 1	What is Colour?	1
Chapter 2	Biological Colourations in Living Organisms	7
Chapter 3	Functional Anatomy of Colour Vision Across Species	21
Chapter 4	Comparative Psychophysics and the Biological Role of Colour Perception in Animals	29
Chapter 5	Colour Perception in Man	41
References		49
Index		59

PREFACE

This new book first examines how colour is produced by animal species including insects, fish, birds and mammals. It then explores psychophysical studies on colour perception in both animals and man. These demonstrate that colour is detected and processed in animals and help to understand the complex mechanisms involved. It examines the biological role of animal colourations for defense, aggression and, above all, in the search for a sexual partner to propagate the species. Finally for man, colour perception is important since it contributes either to rewarding psychological sensations of warmth, comfort and safety or to aversive sensations of coldness and discomfort, sensations which can have a strong influence on our daily social interactions.

AUTHORS' BIOGRAPHICAL DETAILS

Birgitta Dresp-Langley, born in Berlin, Germany, obtained a first degree in psychology followed by a doctorate in cognitive science at Paris University. She obtained a tenured research position with the French state research organisation, the CNRS, employing psychophysical approaches to study visual perception in man, first in Paris and then in Strasbourg. She moved to Montpellier in 2003 where she still works on cognitive processes, in particular those related to the visual system. She combines her expert knowledge of colour perception and her artistic talent as an abstract expressionist painter.

Keith Langley was born in southern England near London and obtained a BSc in chemistry at Birmingham University and a PhD at the Institute of Cancer Research in London. He worked as a university lecturer at the Institute of Neurology in London for more than ten years, acquiring expertise in cytological approaches, before moving to Strasbourg in 1978, where he obtained a research post with the French state medical research organisation, INSERM. He worked at the Centre de Neurochimie in Strasbourg in the field of neurobiology for over twenty years, specializing in electron microscopy to relate structure to function in nervous tissue. He then moved to Montpellier, finishing his research career at the Institute of Neuroscience there. Birgitta Dresp-Langley and Keith Langley are married and currently live in Montpellier.

SUMMARY

There is no colour without light, nor is there colour perception without a sensory organ and a brain to process visual input. We first review in this short book how the colour of objects is produced. Most commonly, this depends on so-called pigments, the molecular nature of which provokes strong absorption of part of the incident light falling on an object. However, colour can also be produced by optical phenomena such as refraction, dispersion, interference or diffraction from ordered structures within objects. A wide variety of photonic microstructures are known in the living world and we will describe specific examples in mammals, birds, fish and insects. Some of these structures reflect light in the near ultraviolet spectral region, which is particularly pertinent for certain birds, insects and fish that are sensitive to these wavelengths. A detailed account of a particularly elaborate structure present in the king penguin beak will be given to illustrate the extent to which evolutionary pressure leads to the elaboration of such structures to satisfy specific needs of birds or animals. Colour can also be produced by bioluminescence i.e. by biochemical processes which generate coloured light within living organisms. Subsequently, the perception of colour in man and animals and its biological significance is dealt with. For man, this includes a discussion of the symbolic meaning of different colours. In many species, especially birds, the colours of plumage and parts of the skin have an important survival function. Such biological colourations may fulfil the role of ornaments that determine mate choice and reproduction of the species, or signal good health, and can help individuals to secure and maintain territorial dominance. Colour perception may also have an underlying survival function in man, but more complex explanations are needed to relate perception to such a function. The colour of an object in the visual field is known to determine the way in which humans perceive relations between objects and their background, particularly which objects appear nearer. This suggests that colour perception is important in processing information about the physical structure of the world. The colour red plays an important role in this process, since it drives mechanisms of visual selection which attract attention to, or away from, objects in the visual field. Psychophysical studies of colour perception in both animals and man help to understand these complex processes. Finally, colour perception in man may contribute either to rewarding psychological sensations of warmth, comfort and safety or to aversive sensations of coldness and discomfort, sensations which can strongly influence individuals in their daily social interactions.

Chapter 1

WHAT IS COLOUR?

Without light there is no colour and even when the level of incident light is too low, man cannot detect it. Only when incident light is sufficiently bright is it possible for man to process this visual input and distinguish different colours in his environment. Thus, colour is a percept resulting from a response of the brain to data received by the visual system. Objects emit light of various wavelengths and these can be detected and analyzed precisely by inanimate machines. Our brain, however, processes signals produced by wavelength mixtures, enabling us to perceive them as a phenomenon we call colour. The perception of colour by a living organism thus requires both a sensory organ receiving external signals and a brain to process these signals and transform them into meaningful representations.

How then can colour be defined? Colour theories have attempted to clarify what we mean by colour, how colours can be ordered, related to each other, and transformed to become new colours. A dictionary may define colour generally, as the aspect of any object that may be described in terms of hue, lightness and saturation, but this tells us little about the true nature of colour. In Ancient Greece, Aristotle developed the first known theory of colour, considering that all colours were a mixture of white and black light. He postulated that the gods sent down colour from the heavens as celestial rays and identified four colours corresponding to the four elements, earth, wind, fire, and water, a view essentially unchanged for many centuries. It was not until the optical experiments performed by Sir Isaac Newton in the mid 1660s when, using a glass prism, he demonstrated for the first time that white light could be split up into all the colours of the rainbow extending from red through to violet, each "refracted" by different amounts (figure 1). He also demonstrated that white light could be recreated by combining these different spectral hues using a second prism. Newton first used the word "spectrum" for this array of colours and recognized that it constitutes a range of energies, of a continuous electromagnetic spectrum, differing by their wavelength. He chose to name seven different hues: red, orange, yellow, green, blue, indigo and violet (figure 1). Each time we see a rainbow, we observe Nature's own confirmation of Newton's experiment, with white light split up not by a glass prism but by countless numbers of rain droplets or ice crystals. While Newton realized that the light rays, or electromagnetic waves, of the spectrum were not, strictly speaking, coloured as such and that "coloured light" does not exist, he recognized these rays had the power to stir up specific sensations of colour, and that colour, like beauty, is "in the eye of the beholder". Typical values of the wavelength of different colours of the visible spectrum are given in table 1. Thus, red is defined as a colour, perceived by man, corresponding to a wavelength of about 650-700 nm. The colour green corresponds to a wavelength of about 550nm, and blue to about 450nm. Several attempts to specify and classify colour further have been made since Newton's invention of the colour circle, which tried to simplify our understanding of how colours could be combined to produce others. Early theories of colour were pure speculation but, with more and more data available, it became eventually possible to explain more and more precisely how the brain processes colour. During the eighteenth century, it was shown that any colour could be obtained by mixing the right proportions of light of three wavelengths, provided they were far enough apart, a principle termed trichromacy. An important advance to explain trichromacy was made in 1801 by the English physician Thomas Young, who suggested that at each point in the retina there must exist at least three small, light-sensitive structures coding for the colours red, green, and violet. His theory was adopted and championed by Hermann von Helmholtz and came to be known as the Young-Helmholtz theory, but its basis was only confirmed over a century and a half later by two independent research groups working in the early 1960's (Brown and Wald, 1964; Marks, Dobelle and MacNichol Jr., 1964). These studies examined microscopically the abilities of single light sensitive receptors, so-called "cones", to absorb light of different wavelengths. Three, and only three, different cone types were identified.

**Table 1. Typical wavelength values of
different colours in the visible spectrum**

Colour	Wavelength (nm)
Red (limit)	700
Red	650
Orange	600
Yellow	580
Green	550
Cyan	500
Blue	450
Violet (limit)	400

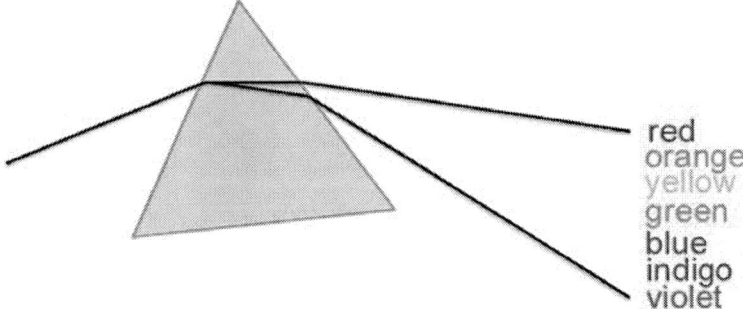

Figure 1. Schematic representation of Newton's prism experiment in which white light is split up into a spectrum of colours, when passed through a glass prism. Different colours have different refractive indices in glass which cause them to be bent to different degrees.

In the second half of the nineteenth century, the German physiologist Ewald Hering (1834-1918) devised a credible alternative theory of colour vision (Krech et al, 1982). He explained colour mixing by proposing the existence, in the eye, brain, or both, of three opponent processes, one for the red-green sensation, one for yellow-blue, and a third for black-white. He positioned four primary colours (red, green, blue, and yellow) according to their polarity as poles of two perpendicular axes and suggested that intermediate colours were formed by additive mixing from the primary colours. The modern Swedish Natural Colour System (NCS) is based on this colour theory. In contrast with this system, the CIE (Commission Internationale de l'Eclairage) established in 1931 an international standard of colour measurement by creating the CIE chromaticity chart which is a graphical representation of colours based on the so-called tristimulus system, which itself is based on the principle that under standardized lighting conditions any colour can be matched to a specific mixture of three primary components of light, the three primary colours; red, green and blue. Such charts graphically represented colours within a triangle, a representation which had already been proposed in 1872 by the Scottish physicist James Maxwell, with the primary colours, at the corners and white at the centre. Maxwell suggested that all known colours could be located within such a triangle. Moving along its edges, red changes to orange, then to yellow, and finally to green; green changes to blue; blue changes to violet, to purple, and back to red. Moving from the edge towards the centre of the triangle, the brilliance of each primary colour is lost in a transition from full saturation at the edge to white at the centre. Maxwell established a system of analyzing any colour, which is identified by the geometric coordinates of any point within the triangle and defined by the quantities of the primary colours it contains, measured by the distance from each primary colour. This system was revised in 1960 and again in 1976 in an attempt to provide an apparently more uniform colour spacing for colours at about the same luminance. Modern chromaticity charts still represent colours within a triangle, although its shape and orientation are very different from the 1931 version. Such representations are currently used to measure and quantify the colours produced on computer screens.

Colour, strictly speaking, refers to the spectral qualities of the light emitted or reflected from an object or living creature. It can thus be defined and measured precisely with spectrophotometers, which analyze the amount of energy present at each spectral wavelength. With such instruments daylight can be shown to be composed of a continuous range of colours corresponding to different wavelengths of unequal intensities, while tungsten light has a predominant red component and very little blue. The spectral reflectance of an artist's pigment in general has a single broad peak the maximum wavelength of which corresponds to its so-called colour.

The term colour should, however, be distinguished from what we understand as biological colouration, which may be defined as the appearance of a living organism determined by the quality and quantity of light reflected by or emitted from its surface. The concept of biological colouration differs from that of colour since it depends on several other factors, including the relative location of coloured areas on the organism, the quality and intensity of light falling on it, the shape, posture and movement of the organism presenting the colouration and, of course, the visual capabilities of the organism looking at it. Since the visual system, i.e. the light sensitive organ and the way in which visual signals are processed by the nervous system vision, is not identical in different animal species including man, fish or birds, colour perception has to be viewed as a species specific process.

Chapter 2

BIOLOGICAL COLOURATIONS IN LIVING ORGANISMS

There are three ways in which an organism may generate colouration: an organism may contain in its surface layers molecules referred to as pigments, which are molecules comprising specific chemical structures that endow them with the capacity of absorbing a proportion of incident light falling on them. Alternatively, colourations may be produced by particular physical structures close to the organism's surface which interact with the incident light so that the reflected light is different. A third way in which an organism may generate colourations is with biochemical processes, using chemical energy to produce endogenous light, which in general is coloured. Each of these will be discussed in greater detail, with pertinent examples from nature and the animal world.

PIGMENT BASED COLOURATIONS

The different groups of pigments found in the natural world impart widely different colours to plants and animals, varying from the brilliant colours of certain fungi and flowers, to the various browns, reds and greens of many organisms. Animal pigments are located in the skin and skin derivatives, which include hair in mammals, feathers in birds, scales in turtles and tortoises, cuticles and shells of many invertebrates. Pigment containing cells or chromatophores, which are derivatives of the neural crest, comprise melanophores, which contain black melanine pigments, erythrophores, which contain orange/red carotenoids and xanthophores, which contain yellow pteridines or xanthopteridines (Bagnara,

1966). A blue pigment containing chromatophore, the cyanophore, has recently been described (Fujii, 2000), but appears to be relatively rare. In mammals and birds, in contrast to cold-blooded animals, melanin pigments are synthesized in melanocytes, the biology and physiology of which differ considerably from that of melanophores. Some chromatophores contain vesicles containing more than a single pigment, such as pteridines and carotenoids for example, and the overall colouration generated then depends on their relative amounts (Matsumoto, 1965). Sudden changes in colouration in certain species may result from stress and/or changes in mood or temperature and thus be under hormonal or nervous control, or a combination of both. This may induce changes in size of chromatophores, migration of pigments within, or a reorientation of microstructures.

How do pigments within cells generate colourations? Since energy in a closed system is conserved and can be neither created nor destroyed, when a photon of light is absorbed by matter constituted of molecules, its energy is absorbed by the matter. It is then either converted into heat, or emitted in the form of a photon of light. The energy of such emitted photons is determined by the energy states permitted for the molecule in question, the difference between such states being explained by quantum theory. Complex organic molecules containing multiple double bonds with the resulting excess of orbiting electrons have the type of structures with vibrational or rotational energy, typical of pigments. The more complex the molecule, the more possibilities there are for different energy levels to exist, but even very simple molecules, like water, are subject to this phenomenon and are thus capable of being coloured. In the solid state, the bonds between hydrogen and oxygen are stronger than in the liquid state, which results in adsorption of white incident light at the red end of the spectrum thus producing the pale blue colour characteristic of bulk ice. Natural pigments are complex organic compounds which contain a "colour bearing" group or "chromophore", often made up of a chain of atoms joined by alternating single and double bonds. If these chains are long enough, they can absorb in the visible spectrum and emit in the visible region.

Two broad classes of natural pigments can be distinguished on the basis of their chemical composition: those containing nitrogen and those without it. The most important non-nitrogenous pigments in animal colourations are the carotenoids. Others, such as the napthoquinones, anthraquinones and flavanoids, which, as is also the case for carotenoids, are only synthesized in plants, do not contribute significantly to animal colouration. The most important nitrogenous pigments in animal colouration are the melanin polymers, which in contrast to carotenoids are synthesized by animals. Other pigments of this general class include the porphyrins (pyrole containing molecules) and their derivatives, which form the red or green compounds in blood, and the green chlorophylls of plants.

The carotenoids are constituted of long chain conjugated systems which can absorb violet or blue light, resulting in their yellow, orange or red appearance. Carotenoids contribute to the yellow, orange, and red colours of the skin, shell, or exoskeleton of aquatic and other animals. They are ubiquitous in living organisms and are the most widespread pigments found in nature; they are synthesized by bacteria, fungi, algae and many plants (Shahidi et al., 1998). They must be absorbed from these organisms to produce animal colouration. A typical example is the flamingo (figure 2), the characteristic pink colour of which is derived from the beta carotene in their diet of shrimps and blue-green algae, which is subsequently transported into the erythrophores (Bagnara, 1998). In captivity, flamingos are often given a diet supplemented with canthaxanthin, a ubiquitous carotenoid pigment first isolated in edible mushrooms and also found in green algae, bacteria, crustaceans, and certain fish. This makes them an even greater attraction to zoo visitors. Canthaxanthin is also fed to farmed salmon to give them a more attractive pink colouration. Carotenoids are well documented in both feathers and bills of certain birds (McGraw and Nogare, 2004; Peters et al., 2004). Their quantity in the plumage of an individual is strongly influenced by diet and may thus reflect the bird's general condition (Price, 2006).

Figure 2. Pink flamingos in captivity, illustrating the carotenoid based pink colouration of their feathers and skin. This is enhanced by feeding these birds with a carotenoid enriched diet.

A second major group of animal pigments are the indoles, which comprise the melanins. They are present in the melanophores of lower animals and in the equivalent higher vertebrate pigment cells, the melanocytes. These pigments, which include eumelanin, a complex molecule synthesized from tyrosine with dihydroxyindole-2-carboxylic acid and several pyrole rings, and phaeomelanine, give rise to buff, red-brown, brown and black colourations in the feathers of birds, mammalian hair, eyes, and in the skin and scales of many fish species, amphibians and reptiles, in squid and octopus ink, and in various invertebrate tissues (Ito and Wakamatsu, 2003). These pigments are present in granules throughout melanocyte cytoplasm. Although nitrogenous pigments, such as melanins, are predominant in marine organisms, pigments belonging to all the major structural classes of natural products are also found (Bandaranayake, 2006). Colourations produced by melanins are affected by different levels of traces of copper, light hair containing less than dark hair, for example. Melanins, like carotenoids, are common in bird feathers, but melanin patterns in birds seem to be less affected by diet than carotenoids. The intensity of melanization after moult is affected by social interactions during the moult and by raising birds in humid conditions. Hormonal manipulations can have dramatic effects on both the kinds of melanin produced (eumelanin or phaeomelanin) and the patterns they form (Price, 2006). The melanin pigments in fish skin are dependent on breeding conditions (Seikai et al., 1987). It has recently been suggested that carotenoids may be a condition-dependent trait, whereas melanin-based colouration is not, a difference that may be highly relevant when studying the evolution of multiple mating preferences. The expression of these two traits may be regulated by different mechanisms (Senar et al., 2003).

A pigment, the chemical structure of which has only recently been elucidated (McGraw and Nogare, 2004), gives rise to the red colouration of parrot feathers, to which it owes its name, psittacofulvin (literally in Greek: parrot colour). The chemical structure of the group is based on long unsaturated carbon chains with an oxygen terminal. Unlike carotenoids, which are absorbed from food, these are synthesized by parrots, and apparently only by them, which means that the red colour of parrot feathers does not depend on diet and will not fade when this is lacking in carotenoids. The green colours of parrot feathers result from a combination of blue structural colours (see later) and a yellow pigment which could be related to the psittacofulvins.

Other pigments have important biological activities in certain organisms. Pteridines, which are molecules composed of fused pyrimidine and pyrazine rings containing a wide variety of substitutions, synthesized from guanosine triphosphate (GTP), are the red-yellow pigments of composite eyes of arthropods

They were first discovered in butterfly wings (hence their name from Greek pteron; wing). Other purine derivatives, present in leucophores, are not pigments, but form the white crystals producing the structural colourations (see below) of amphibians and lizards and cephalopods, also responsible for the whitish skin colourations on the undersides of many fish.

Another less common pigment of the quinine group, the naphthoquinones, which can be red, purple or green are present in shells of echinoids or sea urchins. The yellow or orange red flavonoids are present in plants but are rare in the animal kingdom, although a sub class has been found in the yellow wing colouration of some species of butterfly. Porphyrins represent an important class of pigments in the plant kingdom, although they are present in the skins or shells of invertebrates, forming the green chlorophyls of higher plants and the hemoglobins of animal blood.

STRUCTURE BASED COLOURATIONS

Colour can be produced by an object purely in accordance with the laws of optical physics if it contains sub-microscopic, so-called photonic, structures that are able to modify incident light, by either specifically absorbing or reinforcing certain of its component colours. This manipulation of incident light originates from the basic optical phenomena of thin-film interference, multilayer interference, and reflection or diffraction grating effects, (Dyck, 1976; Parker, 1998; Vukusic et al., 2001; Kinoshita and Yoshioka, 2005). Light scattering similar to that occurring in certain types of structural colours present in living organisms has been reproduced in purely artificial systems consisting of polystyrene microspheres (Guillaumée et al., 2008). Artificial grating structures can be fabricated with electron-beam lithography, which can produce iridescent blue colours by the same process as that which occurs in the wings of certain butterflies. Nature has evolved an extraordinary diversity of highly specialized structures capable of reflecting different colours or ultraviolet light, UV, (Parker, 2000; Vukusic and Sambles, 2003), which have been well documented in insects, fish, butterflies, birds and plants. In such photonic structures, white light can be fractionated and thus modified by interference after reflection from successive ultra-thin layers, often producing iridescent effects where the perception of colour varies with viewing angle. The wavelength of the reflected light is governed by physical laws, such as Bragg's law, and depends on both the angle and the distance between the parallel successive reflective layers, a phenomenon termed coherent scattering. A different type of scattering, Rayleigh scattering, termed non coherent and similar to the effect first discovered by the British physicist Tyndall in the 19th century, is due to the preferential reflection of light of shorter wavelengths (blue) by finely dispersed particles in a material, while longer wavelengths (red and yellow) pass through. This is why the sky appears blue on a sunny day and for more than a century the blue colouration of living organisms was thought to be due to Rayleigh scattering, but many studies have since categorically refuted this interpretation. Raman was the first to question it (Raman, 1935), postulating that the colour was due to the interference of light, a study ignored until 1971 when Jan Dyck, a Danish scientist, concluded that the blue and blue-green colours produced by the spongy structure of feather barbs were not due to Rayleigh scattering, but to the interference of light by backscattering from numerous hollow, randomly oriented keratin cylinders (Dyck, 1971). In 1998 and 1999, Prum and co-workers (Prum, Torres, Williamson and Dyck, 1998; Prum, Torres, Williamson and Dyck, 1999) confirmed that feathers look blue because of differences in the distances travelled by light waves reflected off of each successive layer in the spongy keratinous layer in the barbs. Thus, in fishes, birds and mammals including humans, blue is almost always a structural colour based on coherent scattering, while in lower vertebrates, blue is produced from crystalline platelets termed schemochromes, which are constituted of guanine in iridophore cells. Their orientation determines the brightness and iridescence of the colour that is perceived (Taylor, 1969; Morrison, 1995). A rare exception to this was recently found in certain fish, the blue colour of which is due to a blue pigment of unknown chemical structure present in the vesicles of cyanophores (Fujii, 2000). The leucophore, a structural colour producing cell found in many fish species, has crystalline purine reflectors, which produce a dazzling white (Fujii, 2000). In amphibians, reptiles and birds, the scatter of blue wavelengths, together with the presence of yellow pigmentation, is fundamental for the expression of green colourations (Bagnara, Fernandez and Fujii, 2007; Fox, 1979). A detailed description of all the different types of structures capable of generating biologically relevant colourations found in the animal kingdom would go beyond the scope of this chapter. However, general examples will be described and a detailed account will be given of a particularly relevant, highly complex UV-reflecting structure, recently found in the King Penguin beak.

Considerable interest has been devoted to the structural colourations of butterfly wings and detailed information has been reported on the nature of these structures. All butterfly and moth scales and bristles are made of non-living insect cuticle, each produced from a single epithelial cell. Even though all cuticular patterns in insects have common basic elements, some are highly specialized, with stacks of thin-films, lattices, or other minute structures which interact with light to produce structural colours (Ghiradella, 1994). The numerous studies on the structural colourations of butterfly wings are now in general agreement that multilayer reflectors are commonly at their origin and these are recognised to produce the most intense structural colours (Parker, McPhedran, McKenzie, Botten and Nicorovici, 2001). Multilayer reflectors with similar optical properties are common in other insects and have even been found in fossils of an extinct species of beetle (Parker and McKenzie, 2003). Despite their anatomical diversity at the nanoscale level, all structurally coloured butterfly scales share a single fundamental physical colour production mechanism - coherent scattering (Prum, Quinn and Torres, 2006). The blue colour of the wings of the Morpho butterfly is generated by multilayer microstructures with as many as 24 layers (Wong et al., 2003; Kinoshita, Yoshioka and Kawagoe, 2002; Vértesy et al., 2006), often producing a marked iridescent effect (Vukusic et al., 2002). Some butterflies reflect ultraviolet light, a phenomenon often reserved to the male of the species which suggest a role in sexual communication (Ghiradella et al., 1972). Similar optical interference produces the bright green dorsal iridescence in species as diverse as insects (Vukusic, Wootton and Sambles, 2004) and molluscs (Brink, van der Berg and Botha, 2002).

Although structural colourations have been found in many different animal species, their biological significance has been investigated mainly in birds (for reviews see Auber, 1957; Dyck, 1976). Several physical mechanisms have been evoked, but avian plumage structural colour is now considered to result from coherent light scattering, either from a spongy keratin-air matrix (Prum et al., 1998) or from photonic crystals constituted of melanin rods within a keratin-air matrix (Zi et al., 2003). This contrasts with the structural skin colour, produced by coherent light scattering from ordered collagen arrays in more than 2.5% of all avian species (Prum and Torres, 2003; 2004), and which can also produce UV reflectance in the facial skin of certain birds (Prum, Torres, Kovach, Williamson and Goodman, 1999). The display of the male Indian Peafowl (peacock) is remarkable enough to attract man, even though it has evolved this highly specialized visually striking colouration by evolutionary pressure only to attract peahens (figure 3). This bird's iridescent blue-green and green coloured plumage, which shimmers and changes with viewing angle as in many bird feathers, results from coherent scattering from periodic nanostructures made up of the melanin layers in the barbules. Different colours correspond to different distances between the periodic structures. Brown feathers are produced by a mixture of red and blue, produced by varying the lattice constant and the number of periods (Zi et al., 2003; Li et al., 2005). Some birds employ keratin fibres to produce structural colours, as in the green and purple barbules of pigeon feathers, figure 4 (Yin et al.,

2006). Structural colourations can, in addition, combine with pigment based colourations in bird feathers (Osorio and Ham, 2002; Shawkey and Hill, 2005).

Figure 3. The dramatic display of the tail feathers of the peacock. These colouration patterns, which in addition to their colour are iridescent, illustrate the degree to which evolutionary pressure can exert to produce immensely complex colourations by structural mechanisms.

Figure 4. The green and purple iridescent colourations of pigeon neck feathers. Although these are far less complex and more modest than those of the peacock, they are caused principally by the same mechanism, i.e. by structural colours.

Since most birds detect near-UV light and pigments emitting in the UV region are hitherto unknown, photonic structures reflecting in the UV range have aroused considerable interest in recent years, particularly with regard to their potential biological function (Finger and Burkhardt, 1994; Mougeot et al., 2006). Such structures were first discovered in avian plumage (Hausmann et al., 2003; Prum Torres, Williamson and Dyck 1999), but are also present in other avian tissues, including skin (Mougeot et al., 2005; Prum, Torres, Kovach, Williamson and Goodman, 1999; Prum and Torres, 2003), mouth tissue (Hunt et al., 2003) and comb tissue (Mougeot et al., 2005). The King Penguin beak provides a particularly interesting example of UV structural colour. This bird has several coloured ornaments including the carotenoid based yellow breast feathers, but only the beak horn, which is perceived by the human visual system as yellow-orange coloured, with a more or less pronounced pinkish-violet tint (figure 5), reflects as shown by reflectance spectrophotometry, in the near UV region (figure 6), with an average peak around 370nm. A second, broader peak typical of carotenoids is present in the yellow-orange-red region of the visible spectrum.

Figure 5. King Penguins in nuptial parades on the beach in Crozet island in the sub-Antarctic. Note the way in which they flaunt their beaks in face to face encounters with potential partners.

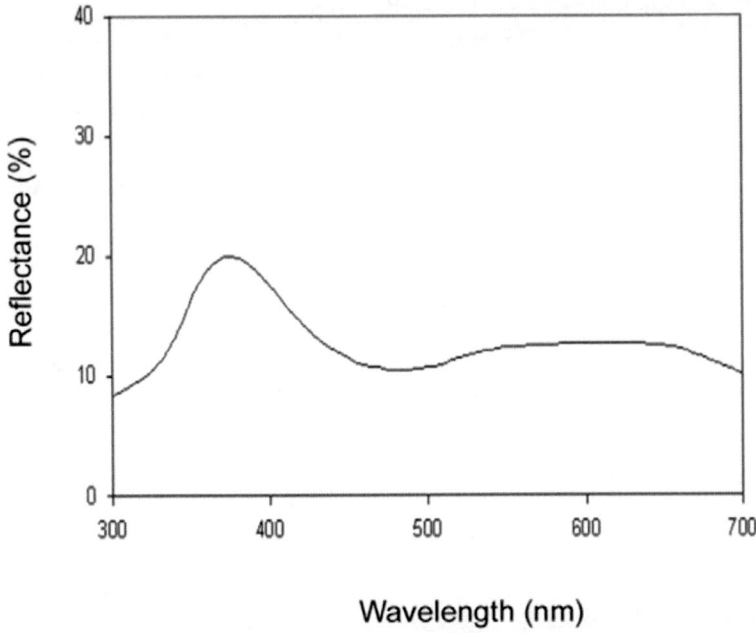

Figure 6. Reflectance spectrum from the King Penguin beak horn, showing a pronounced peak around 370nm in the ultraviolet range and a very broad peak due to carotenoids in the visible region.

We (Dresp and Langley, 2006) demonstrated that only structures situated within the upper region of the beak horn, comprising a corny layer (*stratum corneum*) of this specialized skin tissue, reflect UV. This region is made up of interconnected microstructures of a multiply folded membrane doublet that produces up to forty quasi-parallel layers in individual microstructures, formed from intergititated cell plasma membranes of adjacent keratinocytes (figures 7, 8). Calculations using lattice dimensions measured by transmission electron microscopy accurately predicted the wavelength experimentally measured of the near ultraviolet reflectance from these multiple layered microstructures. These calculations were based on Bragg's law (Bragg and Bragg, 1915):

$$\theta_{max} = n2d\ sin\theta$$

where θ_{max} is the peak wavelength of reflected light, n is the average refractive index of the tissue, d is the separation of the layers (lattice dimension) and θ the angle of incidence of the light (here 90°). As in all animals, the quality of the structural colour in a bird may reflect its individual hormonal status or health and, therefore, its intra-specific ability to compete and, therefore, to survive.

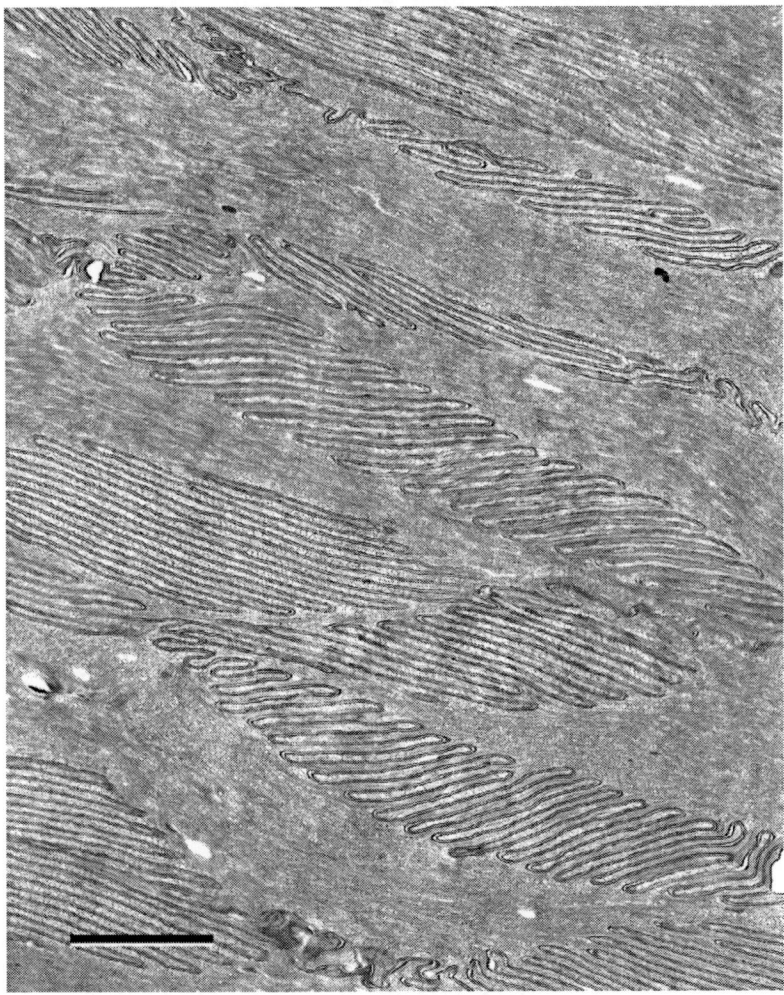

Figure 7. Low power transmission electron micrograph of the upper region of the King Penguin beak horn, which is filled with interconnected microstructures constituted of multiple layers of folded membranes, each of which is formed from a doublet. The microstructures vary considerably in shape and size. Irregular bundles of intermediate filaments fill the space between them. Scale bar 2.0 µm.

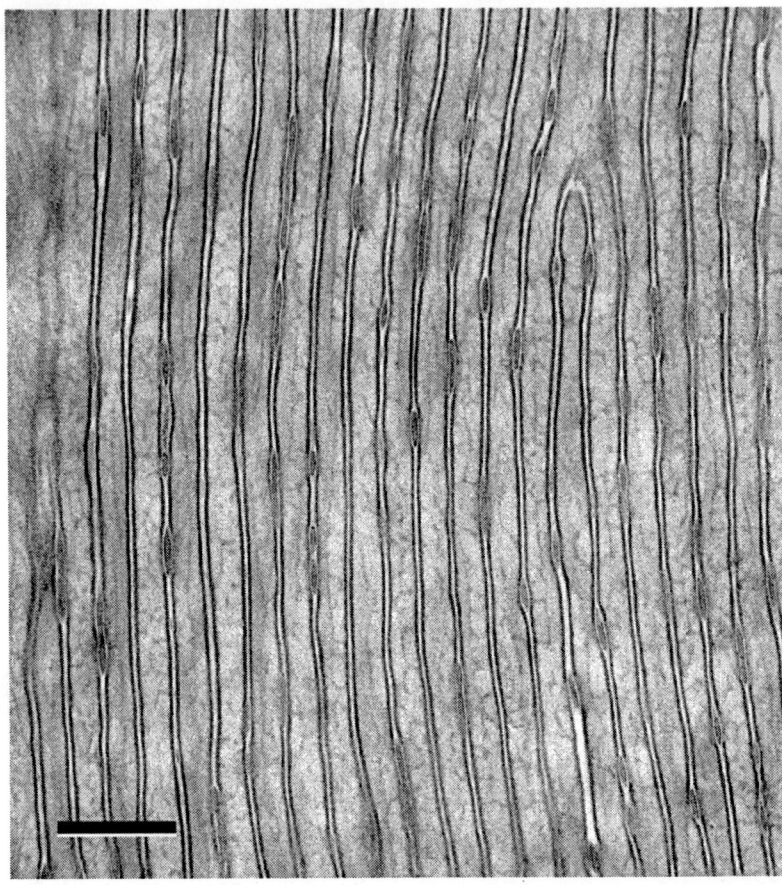

Figure 8. Higher magnification electron micrograph of part of a microstructure present in the upper region of the King Penguin beak horn, showing parallel reflecting multilayers composed of folded cell membranes compressed together. Note also that one of the folds is shorter than the others. Scale bar 0.5 µm.

BIOLUMINESCENCE

Although we live in a world where the most important light source is the sun, man learned early to produce artificial sources of light, which enabled him to cope in the absence of natural light or under low lighting conditions. Light can also be produced endogenously by certain living organisms, through biochemical processes occurring in light producing organs, so-called photophores. This form of light production, known as bioluminescence, is more often present in organisms which live in low light habitats, such as deep water fishes or nocturnal species of insects. Bioluminescence is mainly a marine phenomenon not found in freshwater. On land, it is seen only in a few species of fungi and insects. There are no luminous flowering plants, birds, reptiles, amphibians or mammals. Bioluminescence is phylogenetically diverse, occurring in many different groups, including bacteria, fungi, dinoflagellates (algae) coelenterates (jellyfish), annelids (sea worms) molluscs (squid and clams), crustaceans (shrimps), insects (firefly), echinoderms (brittle stars) and many bony and cartilaginous fish (Herring, 1987). Bioluminescence is most prevalent at mid-ocean depths, where some daytime illumination penetrates; then surprisingly bioluminescence may occur in over 95% of the individuals. Above and below mid-ocean depths, luminescence occurs in less than 10% of all individuals and species. Among the coastal species, less than 2% are bioluminescent.

Bioluminescence is a form of "cold light emission", less than 20% of the light generating thermal radiation (Hastings, 1983). It is produced by the interaction of a group of substances known collectively as luciferins and the enzyme luciferase, which catalyzes its combination with oxygen to form an oxyluciferin in an electronically excited state, which quickly decays, emitting a photon as it does. Since bioluminescence has evolved independently many times, the genes and proteins involved are unrelated in the different groups of bioluminescent organisms. Thus, each uses its particular organism-specific luciferin and luciferase. Bioluminescence may produce blue, green, yellow, orange, or red light. Land-living organisms tend to produce yellow or green light, and marine organisms blue or green light. In the deep sea, most bioluminescence is blue, the wavelength of light transmitted best by ocean waters. This is no doubt linked to the fact that most marine organisms are sensitive to blue light. Most deep-sea animals can only see blue light, except a few fish, such as the black dragon fish, which appear sensitive to and produce both blue and red light. Because their red bioluminescence is not detected by most deep-sea animals, such fishes may use their red light for intra-specific communication and/or predation.

Non-marine bioluminescence is less widely distributed, but a wider variety in colourations is observed. Famed for their ability to emit light, the glow-worm was aptly described by Wordsworth as "the earth-born star" in his poem 'The Pilgrim's Dream' (*circa* 1820). Bioluminescent fire-flies, like the glow-worm, belong to the same Lampyridae family of beetles. Some bioluminescent organisms, such as certain fungi and bacteria, emit light continuously, while dinoflagellates, a group of marine algae, produce light only when disturbed.

Chapter 3

FUNCTIONAL ANATOMY OF COLOUR VISION ACROSS SPECIES

Colour perception is a result of evolutionary pressure and the ability to see colour has evolved, to a greater or lesser extent, in many different species. To be able to distinguish certain colourations from others plays an important part in the processes which ensure survival. Such ability is determined by the anatomy and functional development of the visual system a given species is equipped with.

In the animal world, at least forty different types of visual systems exist, the simplest just able to differentiate light from dark, while sophisticated ones can distinguish both shapes and colours, the most complex belonging to the mantis shrimp, a predatory crustacean living on the ocean floor. There is an enormous diversity in both retinal structure and visual neuronal mechanisms across the animal kingdom, with a corresponding diversity in the role of colour vision in animal's perception, behaviour, and interactions with the environment (Land and Nilsson, 2002). With regard to retinal anatomy, the vertebrate retina is "back to front" i.e. the photoreceptors are located behind a layer of neurons, as opposed to the cephalopod retina, in which the photoreceptors are located in front of the processing neurons, which means that cephalopods do not have a blind spot. Moreover, the cephalopod retina is not an outgrowth of the brain, as in vertebrates, illustrating that vertebrate and cephalopod eyes have evolved separately.

COLOUR VISION IN NON-HUMANS

At the start of the 20th century colour vision was thought to be exclusive to man. Now however it is well established that colour vision is widespread in non-human animals, although even among vertebrates, the ability to perceive colour and the spectral range detected varies widely, having evolved several times. Nevertheless, the precise colour perception capabilities of the majority of animal species are still not known, since behavioural or physiological tests for colour vision are not easy to perform. Human colour vision (see later) is made possible by the presence of three types of colour sensitive receptors, a lack of colour receptors (cones) sensitive to red, blue, or green light provoking deficiencies in colour vision or various kinds of colour blindness. In general terms for all living organisms, if they possess only a single retinal pigment then they will only "see" in monochrome and for even limited colour vision at least two types of cones are required. However the number of retinal pigments in animal species is not limited, as in man, to three. Using behavioural studies and spectrophotometric analysis of light absorbed by animal retina, it can be deduced that four exist in approximately thirty species of birds and five in certain butterflies and many more for sea organisms such as the mantis shrimp and the sea manta, a giant ray (Cronin, Caldwell and Marshall, 2001; Marshall, Cronin and Kleinlogel, 2007). While the human eye cannot detect all of the electromagnetic spectrum emitted by the sun, notable UV, which in any case is absorbed by the human cornea, lobsters, gold fish, trout, bees, tortoises, many bird species and rodents such as rats and mice can perceive in this region (Cuthill et al., 2000). At the other end of the spectrum, some species, such as snakes, perceive infra-red but these rays are captured by specialized heat sensitive organs and not by the retina.

Colour vision of non-primate mammals is still a question of on-going research, varying considerably between species and often dependant on the nocturnal or diurnal nature of the animal. Much research is currently devoted to elucidating the capacity of colour perception in many different animal species, but there remains a vast number for which little concrete data exist. While certain mammals, such as the shrew (in spite of having a reputation of poor sight) and certain squirrel species are considered to be trichromatic, in general non-primate mammals are considered to have relatively limited colour vision. The variation in colour vision results from the fact that the capacity to detect colour has evolved more than once, with gene duplication for visual opsin pigments (Bowmaker, 1998). In addition, probably because of their nocturnal ancestry, mammals have rod-dominated retinas and many lack cones with red sensitive pigment and therefore have poorer (only dichromatic) colour vision: some species lack cones completely, such as the guinea pig and are thus totally colour blind. Bulls, like many mammals and in spite of the importance of the colour red in bull-fights, are insensitive to this colour. Similarly, the cat is dichromate with one type of cone sensitive to blue/indigo ca 450nm and another to yellow/green ca 556nm, thus incapable of perceiving red, as is also the case for the ferret (Calderone and Jacobs, 2003). Cats do however have more rods (the highly light sensitive receptors) and less cones, giving good night, albeit monochrome, vision. Behavioural studies performed in the 1970's suggested that rabbits have a rather limited capacity to distinguish certain wavelengths, although they can differentiate between green and blue. Horses also have only dichromatic vision detecting blue and green and the colours based on them, even though the equine eye is the largest of any land mammal (Carroll et al., 2001). The limited colour perception of horses is sometimes taken into consideration when designing obstacles in horse jumping events. Horses, however, have more rods than humans, in addition to twenty fold more rods than cones, giving them better night vision.

A different system has evolved in rats and mice which have excellent night vision, due to a higher number of rods than cones, but see poorly in colour although both are dichromatic. Both rat and mouse cones co-express two photopigments, one sensitive to wavelengths of ca 510nm and one to ca 360 nm i.e. UV. Their visual systems exploit differences in the spectral absorption properties among the cones, enabling them to make certain dichromatic colour discriminations, which is also the case for gerbils (Jacobs and Williams, 2007; Jacobs, Fenwick and Williams 2001; Jacobs, Williams and Fenwick, 2004; Jacobs and Deegan 2nd., 1994). Both diurnal rodents and rodents which live in almost lightless conditions have been found to have similar colour vision (Williams, Calderone and Jacobs, 2005; Jacobs et al., 2003).

Amongst mammals, primates represent exceptions with regard to colour perception. Sound data confirm the long held suspicion that colour vision in primates including humans, apes, and Old World monkeys is better developed than in other mammals: in spite of data lacking on many species, these are all considered to be trichromatic (Jacobs, 1993). Variations nevertheless exist (Jacobs, 1996; Jacobs and Deegan 2nd., 1999). Variations amongst New World monkeys are even greater, some species being trichromatic while others are only dichromatic (Jacobs and Williams, 2006). In addition, evidence predicts that all male New World monkeys are dichromats while, depending on their opsin gene array, individual females can be either dichromatic or trichromatic (Jacobs and Deegan 2nd., 2003; 2005; Rowe and Jacobs, 2004). Some nocturnal species appear to be monochromatic however (Jacobs 1996).

The situation for aquatic mammals is quite different. Many species, including dolphins and seals, and in particular mammals that live in deep water tend to have blue shifted vision compared to that of many terrestrial mammals and are monochromatic (Fasick et al., 1998). This is considered to result from the absence of evolutionary pressure to maintain colour in the dark monochromatic oceanic environment (Newman and Robinson, 2005).

Of species studied so far, the best colour vision appears to be found in lower vertebrates such as certain birds, aquatic creatures, and certain insects including butterflies and honeybees. Birds vary according to species in their capacity to perceive colour. Diurnal birds tend to have increased ultraviolet sensitivity, with far more cones than rods and their cones are sometimes complex, while nocturnal species such as owls tend towards sensitivity in the infrared end of the spectrum with a relatively high proportion of rods and are colour-blind. The most acute avian vision is found in raptors, such as hawks and eagles that rely on their sight to spot prey from altitude. They are bifoviate, increasing the potential number of cones, some of which are double, a phenomenon also observed in fish, amphibians, and reptiles. An eagle has a larger retina than a human with five times more cones and its structure, like that of most birds, is more complex. Many birds have four types of cone making them potentially tetrachromats.

Fish appear to have quite well developed visual systems, comparable in some species to those of birds. Some have photoreceptors with peak sensitivities in the ultraviolet range. This may be because they move about in a blue environment and need to contrast food sources or predators against a blue background. Teleostei, regrouping about 23,600 vertebrate species making up 95% of all known fish species, perceive red, yellow/green/blue, violet and UV up to 365nm. However, since sea water selectively absorbs longer wavelengths, i.e. red light, many fish living below 10 metres see "poorly" in the red region. Nevertheless many reef fish species living at this depth emit red fluorescence the origin of which are guanine crystals and do perceive this colour (Michiels et al., 2008).

Amphibians see fairly well in colour with a maximum day vision principally in yellow and at night in green. Colour vision in reptiles is also well developed with turtles able to distinguish between blue, green and orange and lizards between yellow, red, green and blue. Reptiles have genera that possess rods and four spectral classes of cone each representing one of the five visual pigment families, endowing these species with the potential for tetrachromatic colour vision (Bowmaker, 1998).

Many insects, including butterflies, flies and bees, have colour vision. Butterflies vary widely in their sensitivity to light, and are considered to have the widest visual range of any form of wildlife. The Chinese yellow swallowtail butterfly (Papilio xuthus) has a pentachromatic visual system, i.e., the eyes contain five different types of cones, sensitive to UV, violet, blue, green, and red wavelength peaks. In nature, these butterflies feed on nectar provided by flowers of various colours not only in direct sunlight, but also in shaded places and on cloudy days. The windmill butterfly (Atrophaneura alcinous) has a visual spectral range from at least 400 nm to 700 nm, while the Sara Longwing butterfly (Heliconius sara) has a range from 310nm to 650nm. Mosquitoes don't see yellow but do perceive purple, which is why mosquito traps emit blue to ultraviolet light which attracts these insects. Bees, have complex compound eyes, four visual cells in each ommatidium responding to yellow/green light (530 nm), two responding maximally to blue light (430 nm) and the remaining two responding best to ultraviolet light (340 nm), allowing the honeybee to distinguish colours (except red). However, an additional feature of bee vision is that they are dichroic, i.e. sensitive to polarized light, which would pose a problem in perceiving colour from waxy plant surfaces since this is partially linearly polarized. Bees and many other insects overcome this problem since the majority of their photoreceptors are twisted like a corkscrew which enables them to perceive the same colour in all directions (Wehner and Bernard, 1993).

COLOUR AND THE HUMAN VISUAL SYSTEM

At the present time much is known about the visual system of man and the way man perceives colour, much more than for any other animal species, and although we may not understand all, we do not expect many surprises. However, even in the distant past two major ancient Greek schools attempted to provide a primitive explanation of human vision. The first theory, championed by scholars like Euclid and Ptolemy and their followers, was the "emission theory" which considered that vision was "produced" when rays emanating from the eyes intercepted objects. The second theory advocated by Aristotle, Galen and their followers, considered vision to result from something representative of the object entering the eyes. While it contains essential aspects of present day explanations, light did not play any role in this theory and it remained only a speculation lacking any experimental foundation. In 1021 Ibn al-Haytham (also known as Alhacen or Alhazen), the "father of optics", was the first to reconcile both schools of thought in his Book of Optics. He argued from extensive scientific experimentation that vision is due to light from objects entering the eye and was the first to suggest that visual perception occurs in the brain, rather than the eyes. He also pointed out that personal experience has an effect on what people see and how they see, i.e. that vision and perception are subjective. The anatomical studies of Leonardo da Vinci, 1452-1519, were the first to demonstrate the special optical qualities of the eye and he considerably advanced ideas on vision when he realized that there is a distinction between central ("foveal") and peripheral vision.

Today we know that only part of the colour processing performed by the human visual system takes places within the photoreceptors themselves, or before the signals reach the higher levels of the brain (Cronin and Marshall, 2001). Anatomical, microscopical, physiological and biochemical data provide detailed explanations of how the visual system in humans permits information from the environment to be detected and processed. The first stage consists of forming of an image on a light-sensitive membrane, the retina, at the back of the eye, which is actually part of the brain serving as a transducer to convert patterns of light into neuronal signals. In the adult human eye the retina constitutes 72% of a 22 mm diameter sphere, with a 3 mm^2 area lacking photoreceptors where the optic nerve leaves the eye, often called "the blind spot". In contrast, lateral to this is a region referred to as the fovea, which is most sensitive to light and responsible for our sharp central vision. The photoreceptive cells of the retina, due to the presence of a photosensitive pigment, respond to photons of light by producing neural impulses, hyperpolarizing cell membranes and ultimately producing electrical signals in retinal ganglion cells. These are then processed in a hierarchical fashion by different parts of the brain, from the retina to the lateral geniculate nucleus, to the primary and secondary visual cortex of the brain to form a representation of the external environment in the brain. The retina contains two types of photoreceptors, rods and cones. The rods, responsible for our dark-adapted vision are more numerous, 75-150 million compared with the 7 million cones in man, employ a sensitive photopigment called rhodopsin, with a thousand fold more sensitivity to light than those in cones. Rods adapt slowly to low levels of light and their sensitivity peaks sharply in the blue, i.e. shorter wavelengths compared to daylight vision, accounting for the increased apparent brightness of green leaves in twilight. They are better motion sensors and since they predominate in the peripheral vision, this affords better vision of dimmer objects in peripheral vision. Visual acuity or visual resolution is, however, much better with the cones. Cones in man, divided into "red" cones (64%), "green" cones (32%), and "blue" cones (2%), provide the eye's colour sensitivity, making man trichromatic. The green and red cones are concentrated in a spot known as the macula, the centre of which contains densely packed cones and no rods. The "blue" cones have higher sensitivity than the other cones and are mostly found outside the fovea, leading to some distinctions in the eye's blue perception.

However, how colour may actually be perceived is, as the very early intuitions of Leonardo da Vinci and others suggested, most often determined by complex processes and interactions between brain structures well beyond the receptor level. While humans can name the colour they see, it is impossible to tell which colour a non-human animal actually perceives when it responds selectively to a given colouration in an experimental task, for example. This will most likely always remain a mystery to humans. However, cleverly designed psychophysical studies have allowed a better understanding of how non-human animals react to colourations, and how such reactions may be linked to environmental pressures or constraints. Moreover, recent scientific studies have shown that colour perception in humans is also quite strongly influenced by the immediate visual environment or context as well as by cultural factors.

Chapter 4

COMPARATIVE PSYCHOPHYSICS AND THE BIOLOGICAL ROLE OF COLOUR PERCEPTION IN ANIMALS

Over a century ago workers such as J. Lubbock and K. von Frisch developed behavioural criteria establishing that non-human animals percieve colour. Many animals in most phyla have since then been shown to have colour vision. Colour is used for specific behaviours, such as phototaxis and object recognition, while other behaviours such as motion detection are not dependant on colour. Having established the existence of colour vision, research focussed on the question of how many spectral types of photoreceptors were involved. Recently, data on photoreceptor spectral sensitivities have been combined with behavioural experiments and physiological models to systematically study the next logical question: "what neural interactions underlie colour vision?" Kelber, Vorobyev and Osorio (2003) give an overview of the methods used to study animal colour vision, and discuss how quantitative modelling can suggest how photoreceptor signals are combined and compared to permit the discrimination of biologically relevant stimuli.

Comparative psychophysics is able to address questions about how colour perception influences behaviour in the different species. Does a bull get enraged by the colour of a red cape or by its movements? Do colourations help bees to discriminate flowers which are plump with nectar from others? Do specific colourations help a cat detect prey more rapidly? What do animals see when they detect colour and do they actually perceive colour? While no one knows exactly what animals see or perceive, there are several aspects of these questions that can and indeed have been explored. Firstly, we need to know how their visual systems work from a physiological point of view and also the spectral sensitivities of the photo-pigments, which have been reviewed above. More important, however, is the question of how an animal's brain processes colour signal inputs. Many scientists working in the field think that animals "understand" little of what they see, just as humans rarely "understand" abstract paintings, even though they may respond emotionally to and "appreciate" what they see. Although we know that visual abilities differ among animals, there is an important distinction between detecting light that illuminates the retina, and understanding what is actually "out there".

Since animals cannot answer questions about the colours they perceive, research scientists have had to develop experiments in which animals are trained to make selective behavioural choices on the basis of colour. If an animal's food is always placed under a red square rather than a green square, if both squares are otherwise identical from all points of view and re-positioned randomly over time, and if the animal still and consistently keeps looking under the red square when it is hungry, we may indeed conclude that it is capable of distinguishing what we perceive as red from what we perceive as green. Yet, what we know about colour perception in the animal kingdom pales in comparison to that yet to be discovered.

Thirty years ago virtually everything known about primate colour vision derived from psychophysical studies of normal and colour-defective humans and from physiological investigations of the visual system of the macaque monkey, the most popular human surrogate for this purpose. The years since have witnessed much progress toward the goal of understanding this remarkable feature of primate vision. Among many advances, investigations focused on naturally occurring variations in colour vision in a wide range of nonhuman primate species have proved to be particularly valuable. Results from such studies (Jacobs, 2008) were central to expanding our understanding of the interrelationships between opsin genes, cone photopigments, neural organization, and colour processing. This work also generated valuable insights into the evolution of colour perception. Latanov, Leonova, Evtikhin, and Sokolov (1997) studied colour discrimination using an instrumental learning paradigm in monkeys (Macaque rhesus) and fish (Carpio cyprinus L). Confusion matrices composed of probabilities of instrumental responses were treated by factor analysis. The spherical structure of perceptual colour space revealed in both species was found to be similar to that in humans, corresponding to "red-green," "blue-yellow" and the neuronal channels signalling for "brightness" and "darkness".

The dance of the honeybee has been researched extensively, and we have a relatively good understanding of the colour vision of bees and related insects. Mosquitoes and flies have been studied because of their role in spreading diseases, and it has been shown that they are attracted or repelled by specific surface colours, and by specific coloured sources of light. Interestingly, the surface colours they prefer do not necessarily correlate with the light source colours that attract them. Studying colour perception in birds is also quite challenging. Observations clearly demonstrate that different species are attracted to bird feeders of particular colours, and that changing the colour of ambient light can trigger early breeding, or alter fertility rates, by mimicking the change of seasons. It is as hard for us to imagine how birds perceive colour as it is for a colour blind person to imagine full colour vision: it is simply outside our experience. Some species which we see as having identical male and female plumage differ when viewed by ultraviolet light - a difference which the birds themselves are capable of perceiving. Butterflies also perceive colourations and often identify each other quite easily by their ultraviolet markings. The male and female little sulphur butterflies (*Eurema lisa*) differ only in the ultraviolet region, with males being strongly ultraviolet reflective and females not.

Over many millions of years, sea creatures have developed a range of light reflectance properties. One example is the large variation in the patterns and colours of fish inhabiting the world's coral reefs. Attempts to understand the significance of the colouration have been made, but all too often from the perspective of a human observer. A more ecological approach requires us to consider the visual system of those for whom the colours were intended, namely other sea life. A first step is to understand the sensitivity of reef fish themselves to colour. Physiological data has revealed wavelength-tuned photoreceptors in reef fish, and provided behavioural evidence for their application in colour discrimination. Using classical conditioning, freshly caught damselfish were trained to discriminate coloured patterns for a food reward (Siebeck, Wallis, and Litherland, 2008). Within 3-4 days of capture, the fish selected a target colour on over 75% of trials. Brightness of the distractor and target were systematically varied to confirm that the fish could discriminate stimuli on the basis of chromaticity alone. The study demonstrated that reef fish can learn to perform two-alternative discrimination tasks, and provided the first behavioural evidence that reef fish have colour vision. Fishes such as members of the billfish family also appear to perceive colourations. These highly visual predatory teleosts inhabit the open ocean. Little is known about their visual abilities in detail, but previous studies indicated that these fish were likely to be monochromats. However, there is evidence of two anatomically distinct cone types in billfish. The cells are arranged in a regular mosaic pattern of single and twin cones as in many fishes, and this arrangement suggests that the different cone types also show different spectral sensitivity, which is the basis for colour vision. First measurements using microspectrophotometry (MSP) revealed peak absorption of the rod pigment at 484 nm, indicating that MSP, despite technical difficulties, will be a decisive tool in demonstrating colour vision in these offshore fishes. When hunting, billfish such as the sailfish flash bright blue bars on their sides, which also reflects UV light at 350 nm as revealed by spectrophotometric measurements. Billfish lenses block light of wavelengths below 400 nm, presumably rendering the animal blind to the UV component of its own body colour. Interestingly, at least two billfish predatory species have lenses transmitting light in the UV waveband and are therefore likely to perceive a large fraction of the UV peak found in the blue bar of the sailfish (Fritsches, Partridge, Pettigrew and Marshall, 2000). Wavelength discrimination ability of the goldfish has also been investigated on the basis of a behavioural training technique in the UV spectral range (Fratzer, Dörr and Neumeyer, 1994). First, spectral sensitivity was determined for two fishes to adjust the monochromatic lights (between 334 and 450 nm) to equal subjective brightness. The results of the wavelength discrimination experiment showed that, independent of which wavelength the fish were trained on, the relative choice frequency reached values above 70% only at wavelengths longer than 410 nm. Wavelength discrimination between 344 and 404 nm was not possible. These observations cannot be explained on the basis of the cone sensitivity spectra. Instead, inhibitory interactions, which suppress the short wavelength flanks of the short-, mid-, and long-wavelength sensitive cone types in the UV range have to be proposed. Kitschmann and Neumeyer (2005) demonstrated that goldfish are able to categorize spectral colours after habituation to a specific training wavelength. Subsequently, Poralla and Neumeyer (2006) trained goldfish on more than one wavelength to prevent very accurate learning. In one experiment goldfish were trained on six adjacent wavelengths with equal numbers of rewards, and, thus, equal numbers of learning events. Generalization tests showed that some wavelengths were chosen more often than others, indicating that certain spectral ranges are either more attractive or more easily remembered than others. This is a characteristic of the "focal" colours or centres of colour categories in human colour perception and the findings in goldfish may thus be interpreted accordingly. There appear to be four categories in spectral ranges approximately coinciding with the maximal sensitivities of the four cone types, and three categories in-between. Experiments with two training colours have shown that there seems to be no direct transition between categories analogous to human "green" and "red", but that there is a colour analogous to human "yellow".

However, visual systems sensitive to colour are confronted with the fact that the external stimuli are ambiguous because they are subject to constant variations of luminance and spectral composition. Furthermore, the transmittance of the ocular media, the spectral sensitivity of visual pigments and the ratio of spectral cone types are also variable. This results in a situation where there is no fixed relationship between a stimulus and a colour percept. Colour constancy has been identified as a powerful mechanism to deal with this set of problems, but is active only over a short-term time range. Changes that cover longer periods of time require additional tuning mechanisms at the photoreceptor level or at post-receptor stages of chromatic processing. Wagner and Kröger (2005) used the trichromatic blue acara (Aequidens pulcher, Cichlidae) as a model system and studied retinal physiology and visually evoked behaviour after rearing fish for 1-2 years under various conditions, including near monochromatic light (spectral deprivation) and two intensities of white light (controls). In general, long-term exposure to long wavelength light had lesser effects than light of middle and short wavelengths. Within the cone photoreceptors, spectral deprivation did not change the absorption characteristics of the visual pigments. In contrast, the outer segment length of middle and long-wave-sensitive cones was markedly increased in the blue reared group. Furthermore, in the same group, a loss of 65% short-wave-sensitive cones was observed after 2 years. These changes were interpreted as manifestations of compensatory mechanisms aimed at restoring the balance between the chromatic channels. At the horizontal cellular level, the connectivity between short-wave-sensitive cones and the H2 cone horizontal cells were affected in the blue light group. Responses of H2 horizontal cells to light were also sensitive to spectral deprivation showing a shift of the neutral point towards short wavelengths in the blue reared group. An intensity effect was found in the group reared in bright white light, where the neutral point was more towards longer wavelength than in the dim light group. Like changes in the cones, the reactions of horizontal cells to spectral deprivation in the long wave domain may be considered to be compensatory. The spectral sensitivity of the various experimental groups of blue acara in a visually evoked behaviour task revealed that changes in relative spectral sensitivity were too complex to be explained by a simple extrapolation of adaptive and compensatory processes in the outer retina. The inner retina, and/or the optic tectum appear to be involved here, reacting to changes of the spectral environment. Thus in summary, there appears to be considerable developmental plasticity in the colour vision system of the blue acara, where epigenetic adaptive processes at various levels of the visual system respond to the specific spectral composition of the surroundings and provide a powerful mechanism to ensure functional colour perception in different visual environments. Processes involving an active fine-tuning of photoreceptors and post-receptor processing of chromatic information during ontogenetic development are a general feature of all colour vision systems. These appear to attempt to establish a functional balance between the various chromatic channels. This is likely to be an essential condition for cognitive systems to extract relevant and stable information from unstable and changing environments. Comparisons of functionally important changes at the molecular level in model systems have identified key adaptations driving isolation and speciation. In cichlids, for example, the long wavelength-sensitive opsins appear to play a role in mate choice and variations in colourations of the males, within and among species. To test the hypothesis that the evolution of elaborate colourations in male guppies (*poecilia reticulata*) is also associated with opsin gene diversity, Ward, Churcher, Dick, Laver, Owens, Polack, Ward, Breden and Taylor (2008) sequenced long wavelength-sensitive opsin genes in six species of this family and concluded that enhanced wavelength discrimination may be a possible consequence of opsin gene duplication and divergence. This might have been an evolutionary prerequisite for colour-based sexual selection and may have led to the extraordinary colourations now observed in male guppies and in many other species.

Many living organisms thus clearly appear to detect visual information relative to colour, and are able to perceive differences in colourations. How essential is it for their survival, and is its absence life threatening? Several generalities can be postulated: there exists a strong interdependence between the habitat and/or behaviour patterns of different species and their capacity to perceive colour. For example, flying birds require a comprehensive visual perception in three-dimensional space, while birds that live on seeds and fruits in the forest canopy need to differentiate between green and the colours of their chosen foods. Monkeys need to distinguish between orange fruits and green foliage to find food. Ultraviolet vision as possessed by the kestrel can offer a significant advantage in spotting the traces left by prey, such as urine and faeces. The spectral range of vision of bees and butterflies also extends into the ultraviolet, which could aid them in seeking nectar in flowers they pollinate which have specific ultraviolet patterns. Evolutionary forces have been operational in developing specific characteristics, including particular spectral sensitivities and specific colourations, so that species survive better in a highly competitive environment. Thus, biological colourations may have different functions, but they all represent a means of signalling to members of the same or other species. These functions may be divided into three broad categories - deceptive signalling or camouflage, advertising, and repulsive signalling - which we now discuss in greater detail.

DECEPTIVE SIGNALLING OR CAMOUFLAGE

Deceptive signalling comprises the capacity of organisms for camouflage. Homochromy, which does not necessarily involve changing colour, is a strategy which has evolved in living organisms to enable them to escape predators. It consists of attempting to appear the same colour as the immediate environment. It can be considered to be a form of deceptive colouration, or camouflage. The larger sized species tend not to mimic the colourations of specific objects, but try to blend in with the tones present in their environment. Thus the dark spots of leopards allow them to melt into the brush and stripes of other big cats may easily be confused with surrounding vegetation. Other principles underlying camouflage have been described in addition to such background pattern matching, or *crypsis*. *Crypsis* is, in fact, not an optimal strategy of concealment because edge information remains fully available to a potential predator. Another strategy, called disruptive colouration, has long been assumed more effective. Disruptive colouration is a contrast effect where high contrast elements at edges mask the perception of shape despite the fact that edge elements may be independently visible, a phenomenon potentially useful to zebras. Spectrophotometric studies, combined with data on the perception of shape targets by non-human animals (e.g. Stevens et al., 2006), have provided evidence in support of the idea that disruptive colouration reduces the chances of bird predation of artificially created targets.

Smaller sized species, like fish, reptiles, and amphibians, such as those of tropical rain forests, are able to rapidly change their colourations in response to changes in their environment, having developed complex mechanisms that allow them to deceptively "disappear" by blending in with their surroundings. This is achieved on the basis of a redistribution of the pigments in their chromophores, thus changing their colour. In the most well known example of this type of deceptive signalling, the chameleon, different species can change colours to pink, blue, red, orange, green, black, brown, yellow and turquoise. Two cell types are present in the upper layer of their skin, containing yellow and red pigments respectively, above a layer of iridophores, which can reflect, blue light. A layer of melanophores influences the "lightness" of the reflected light. All these pigment cells can rapidly relocate their pigments and thereby alter the colouration of a chameleon almost instantly. Similar processes occur in other animals and detailed molecular mechanisms of pigment redistribution have been studied in several animal species, especially in fish (Logan et al., 2006), and cephalopods such as calamari, cuttlefish and octopi with bright colours (Cloney and Florey, 1968; Demski, 1992). Rapid colour change in many fish species can be provoked by osmotic changes in the chromatophores (Lythgoe and Shand, 1982), and the regulation mechanisms underlying such changes have been extensively studied (Glaw and Vences, 1994; Valverde et al., 1995; Rodionov et al., 1998; Kashina et al., 2002; Aspengren et al., 2003; Mäthger et al., 2003; Sugden et al., 2004; Snider et al., 2004; Logan et al., 2006*)*.

Cuttlefish are particularly clever masters of disguise, rapidly changing colouration to blend with their backgrounds and thereby seeming to vanish from the scene. Recent studies (Cuthill, 2007) have shown that cuttlefish also use this ability to rapidly change their colouration with exactly the opposite aim, breaking their camouflage to "stick out" in order to direct warning messages at specific predators which are particularly likely to be dissuaded by visual signals. This kind of strategic signalling then falls into the repulsive category, which will be discussed later. Indeed, animals change their colouration for various reasons and not always, as is often assumed, for camouflage. Far more frequently, such changes are a genuine mode of communication within and across species and include warning signals in response to stress, attractive signalling to find a mate (see the next section), and signalling illness or submission to predators (Stuart-Fox and Moussalli, 2008; Porras et al., 2003; Deacon et al., 2003; Fujii, 2000).

Another type of deceptive signalling is employed by some bioluminescent marine species, which employ this not to avoid being detected, but to lure prey. Thus anglerfish use their bioluminescence to attract small animals to within striking distance. The cookiecutter shark uses bioluminescence essentially for camouflage. A small patch on its underbelly remains dark and appears like a small fish to large predators like tuna and mackerel, and when these fish try to catch what they perceive as a "small fish", they are captured by the shark. The marine plankton dinoflagellates display an interesting twist on this mechanism. When a predator of plankton is sensed through motion in the water, the dinoflagellate luminesces and, in turn, attracts even larger predators, which then will consume the smaller would-be predator of the dinoflagellate. This kind of, somewhat risky, selective strategy of deceptive signalling, also called "bluffing", is used by predated animals of various species (Langridge, Broom, and Osorio, 2007).

ADVERTISING AND MATE CHOICE

Animals largely employ their colourations to advertise their presence, either to attract members of their own species or repel those of others. Both colour and ultraviolet reflectance play an important role in attempting to attract a sexual partner as well as communicating worthwhile information in mate choice. This is well documented in fish (Boulcott, Walton and Braithwaite, 2005 ; Rick and

Bakker, 2008b ; Modarressie, Rick and Bakker, 2006; Sköld et al., 2008), where bioluminescence has also been shown to play a significant role (Herring, 2000). Fish, such as sticklebacks and Siamese fighting fish, also employ both visible colour and UV in territorial defence (Rick and Bakker, 2008a). Many fish can display dramatic colour changes during courtship displays related to the degree of sexual arousal of the male, which are produced in the short term by pigment redistribution within melanophores.

Many reports have emphasized the potential biological role of both UV reflectance and fluorescence in avian communication (Andersson and Amundsen, 1997; Hunt et al., 1999, 2001; Örnborg et al., 2002; Siitari et al., 2002; Pearn et al., 2003). Male individuals of various avian species exhibit conspicuous colours on their feathers which are the product of sexual selection driven by mating preferences (Andersson, 1994; Darwin, 1871) and the additional capacity of avian ornaments to reflect UV plays an important role during sexual displays (Hausmann et al., 2003; Hunt et al., 2001; Andersson and Amundsen, 1997; Bennett et al., 1997; Hausmann et al., 2003; Finger, Burkhardt and Dyck, 1992; Pearn, Bennett and Cuthill, 2003; Siitari et al., 2002; Parker, 1995).

The highly coloured ornaments of King Penguins (*Aptenodytes patagonicus*), notably the yellow/orange breast and auricular feathers and the two orange/pink UV reflecting beak horns on each side of the beak the have been suggested to be implicated in mate choice (Dresp et al., 2005). It is significant that during courtship displays King Penguins flaunt their beak ornaments when encountering potential partners (figure 5) in a way similar to the flamboyant display of peacocks when they rapidly shake their feathers to produce a shimmering iridescence colour show with the unique aim of attracting their mate. Why is the King Penguin beak so sexy (Carmichael, 2007)? The fact that the horn is both ultraviolet and also orange-pink in colour increases the signal, as more than a single type of photoreceptor in the observer would be activated and its perception would also be heightened by a contrast effect since the tissue surrounding the horn is black. In addition, the multiplicity of microstructures with slightly different orientations producing the UV reflectance spreads both the wavelength and also the angle over which it is reflected, producing a more easily perceptible signal. This suggested biological function is strengthened by the fact that such UV reflecting ornaments are absent in sexually immature juveniles (figure 9), supporting such a hypothesis (Jouventin et al., 2005; Massaro, Lloyd and Darby, 2003; Dresp, unpublished results). Colour, and particularly colour change, in nuptial behaviour is also widespread among fish and has been shown to be hormone dependent (Sköld et al., 2008).

Figure 9. Adolescent, sexually immature King Penguins in their natural habitat on a beach in Crozet island in the sub-Antarctic. They differ from mature adults by their lack of the distinctive orange-pink ultraviolet reflecting beak horn ornament.

Apart from the function of visually attracting potential mates with coloured ornaments, different colours can also communicate information on general fitness of individuals, in particular whether an individual is well nourished and healthy, although in females this is not always the case, indicating that female ornamentation in certain species has evolved by direct sexual selection on females through male choice (Pärn, Lifjeld and Amundsen, 2005). This would thus provide additional criteria particularly for males in attracting female partners (McGraw et al., 2002). Pigment-based ornaments have repeatedly been shown to be condition dependent, even though carotenoid based and melanin-base pigments are regulated differently, (Senar, Figuerola and Domènech, 2003), but more recently structural coloration has also been demonstrated to be condition dependent (Johnsen et al., 2003). Signalling of condition criteria for mate choice has also been evoked in butterflies (Kemp and Rutowski, 2007).

Bioluminescent light emission is also important for communication during courtship to attract a mate, which has been well documented in insects. This is seen actively in both fireflies and glow-worms, the abdomens of which flash periodically with species-specific temporal patterns to attract mates. Both sexes emit specific patterns in an apparent genuine communication for sexual attraction. After mating the female no longer emits light and lays her eggs. Coded bioluminescent signals for mating have also been well-documented in certain small crustaceans such as ostracods. While pheromones may be used for long-distance communication, bioluminescent is apparently useful at close range for "homing in" on the target.

Con-specific communication between individuals of the same species does not have to be sexual but can serve simply to recognise each other, as is the case for neon fish in the murky shallow waters of the Amazonian rain forest, but also for many fish species that form schools to give them a numerical advantage for survival. In-flight signalling of certain butterfly species has also been documented (Vukusic et al., 2002). In addition, certain fish species use their bioluminescence as a form of "night light" for navigation, but this can be used, in addition to the functions already discussed, for communication with other members of the same species. Moreover, if such bioluminescence is red, communication is "private", since few fish perceive red (Michiels et al., 2008).

REPULSIVE SIGNALLING

As well as enhancing camouflage, improving communication and/or conferring reproductive advantage, body colours may also reflect behavioural strategies, notably those related to territorial defence. This may be to defend nests or spawning sites, but can be maintained throughout the year to defend a territory to ensure an adequate food supply, as is the case for reef fishes. This has been shown to be hormonally influenced (Korzan et al., 2008). When males defend territories they are usually bigger and more colourful than their female counterparts. Colour can also sometimes be a warning signal to animals dangerous or of inedible quality of other organisms, red, black and yellow being commonly employed for this. The black and yellow colouration of bees and wasps represent typical examples. Some animals mimic such warning colours to avoid becoming the prey of others, as examples of selective strategic signalling including "bluffing", mentioned earlier, clearly show.

ADDITIONAL FUNCTIONS

The symbiotic relationship between flowers and butterflies and other pollinators such as bees has evolved so that flowers attract pollinators to feed on their nectar. Unique colourful and/or contrast visual cues, sometimes in the UV range, provide an important ploy used by flowering plants to attract potential pollinators, in the same way that chefs in gastronomic restaurants prepare meals high on visual appeal. The colour of flowers of the horse chestnut tree (*Aesculus hippocastanum*) can change from yellow to red when nectar is no longer produced, i.e. when there is no longer need for pollination: red is not perceived by pollinating bees although their compound eyes are trichromatic being able to see UV (Wakakuwa et al., 2005; Wakakuwa, Stavenga and Arikawa, 2007). Butterflies tend to avoid the colour green when feeding, but are attracted to it during egg laying, since the next generation of caterpillars requires a good source of food after hatching. The green photoreceptors are thus not used when foraging, but the others (blue and UV sensitive) are critical in reducing the time taken to find nourishment (Spaethe, Tautz and Chittka, 2001). Plants that do not depend on insect or bird pollination are unlikely to have showy or scented flowers. The ultraviolet patches on some butterflies are directionally iridescent, so that they appear to flicker in flight. This flickering is thought to have an important role in butterfly behaviour and communication.

Chapter 5

COLOUR PERCEPTION IN MAN

It is clear from the intensive research reviewed above that colour plays a major role in the processes which ensure survival, propagation, and every day behavioural patterns of a large number of non-human living organisms. Colour perception is, indeed, the result of a complex evolutionary process that has produced different colour vision systems in various species across the phylogenetic scale, as we have seen above.

How vital is colour perception for man? Is such ability a mere luxury that makes our lives more enjoyable, as some have suggested? Colour signalling is employed everywhere in the world around us, from traffic signals and red warning lights to the bright and colourful advertisements in shops, streets and on the television screen. Research on colour perception in humans tends to suggest that even in man, though perhaps to a lesser extent than in non-humans, colour perception plays an undeniable functional role and allows humans to better cope with specific environmental constraints. Also, phenomenological observations and psychophysical data have shown that the way in which we perceive a given colour strongly depends on the immediate visual context the colour is embedded in.

CONTEXT EFFECTS IN COLOUR PERCEPTION

Colour perception strongly depends on the immediate visual context around the coloured object, its luminance, and its shape. The colour red of a red square, for example, with a constant chromaticity and luminance, is seen as a darker red when against a background with a brighter colour compared with when it is against a background with a darker colour (figure 10). This well-known observation is called simultaneous colour contrast. Such colour contrast phenomena influence other perceptual processes, such as depth perception and relative impressions of "nearer" and "further away" objects in the visual field (Guibal and Dresp, 2005). The colour red appears to play a particularly important role in such colour based depth perception (see also earlier observations by Bugelski, 1967), and the brain mechanisms potentially underlying this have been linked to both colour stereopsis (Dengler and Nitschke, 1993) and neural interactions in area V4 of the primate visual system (Desimone and Schein, 1987). The functional implications of such a dependency between colour and impressions of near and far may be related to colour based perceptual mechanisms driving attention to detail (Yantis and Jonides, 1991). These may well be equivalent to colour specific perceptual mechanisms in non-human animals, which may help a species detect a prey or predator of a specific colour more rapidly, or may help it assess how far or near it is likely to be.

Figure 10. How a colour is perceived depends on the immediate visual context in which the colour is embedded. The red square on the dark green background shown on top here, for example, is perceived as a brighter than that on the light green background at the bottom. The luminance and chromaticity of all four red squares shown are strictly identical. This phenomenon, also known as subjective simultaneous colour contrast, is accompanied by other visual effects. For example, the red square on the lightest green background here seems nearer to the observer than the red square on the darkest green background.

The colour of what we eat or drink contributes another essential aspect in our lives and has a determining influence on whether we choose or enjoy a particular food or beverage. This is because we have built into our basic experiences the notion that a particular foodstuff has to be a particular colour for it to be edible, i.e. its colour will determine acceptance or rejection of it. This is taken into consideration in preparing food and making it appear more attractive. Thus the colour of salmon in a shop is the first characteristic noted by the consumer. Salmon can be, and often is, rendered pinker by adding carotenoids to food given to fish in fish farms (Shahidi, Metusalach, and Brown, 1998). Likewise margarine (from the Greek *margaron*, pearl white), invented in France in 1869 to replace butter, which was then scarce, is given a yellowish tint to suggest more pleasant sensations than those which would be elicited by spreading white fat on our sandwiches.

Despite the fact that black has become the most fashionable colour to wear worldwide, man has not lost the desire to attract partners by colour. It is clear that man can invent many artificial, creative ways of using colour and resort to sophisticated colourful ornaments to attract others. The use of lip stick, make-up, or colourful sparkling jewellery or underwear by humans has such obvious parallels in the animal world that is almost bound to reflect the ultimate expression of the complex evolutionary process that has led to the emergence of colour perception.

REFERENCES

Andersson S and Amundsen T, (1997) Ultraviolet colour vision and ornamentation in bluethroats. *Proc. R. Soc. Lond., B* 264: 15887-11591

Andersson M, (1994) *Sexual selection.* Princeton University Press, Princeton, New Jersey.

Arikawa K, (2003) "Spectral organization of the eye of a butterfly, Papilio". *J. Comp. Phys.* A 189, 791-800.

Aspengren S, Sköld HN, Quiroga G, Mårtensson L and Wallin M, (2003) Noradrenaline- and melatonin-mediated regulation of pigment aggregation in fish melanophores. *Pigment Cell Res.* 16:59–64

Auber L, (1957) The distribution of structural colors and unusual pigments in the Call Aves. *Ibis.* 99: 463-476

Bagnara JT, Fernandez PJ, Fujii R, (2007) On the blue coloration of vertebrates. *Pigment Cell Res.* 20(1):14-26;

Bagnara JT, (1998) Comparative Anatomy and Physiology of Pigment Cells in Nonmammalian Tissues in The Pigmentary System: *Physiology and Pathophysiology,* Oxford University Press.

Bagnara JT. 1966 Cytology and cytophysiology of non-melanophore pigment cells. *Int. Rev. Cytol.* 20:173–205.

Bandaranayake WM. (2006) The nature and role of pigments of marine invertebrates. *Nat. Prod. Rep.* 23 (2):223-55

Bennett AT, Cuthill IC, Partridge JC and Lunau, K. (1997) Ultraviolet plumage colors predict mate preferences in starlings. *Proc. Natl. Acad. Sci. U.S.A.* 94: 618-21

Boulcott PD, Walton K, Braithwaite VA. (2005) The role of ultraviolet wavelengths in the mate-choice decisions of female three-spined sticklebacks. *J. Exp. Biol.* 208 (Pt 8):1453-8.

References

Bowmaker JK. (1998) Evolution of colour vision in vertebrates. *Eye.* 12 (Pt 3b):541-7

Bragg WH and BraggWL 1915 X-rays and crystal structure. G.Bell, London.

Brink DJ, van der Berg NG, Botha AJ. (2002) Iridescent colors on seashells: an optical and structural investigation of Helcion pruinosus. *Appl. Opt.* 41(4):717-22

Brown PK, Wald G. (1964) Visual pigments in single rods and cones of the human retina. Direct measurements reveal mechanisms of human night and color vision. *Science.* 144:45-52.

Bugelski, BR. (1967) Traffic signals and depth perception. *Science.* 157, 1464-1485.

Calderone JB, Jacobs GH. (2003) Spectral properties and retinal distribution of ferret cones. *Vis. Neurosci.* 20(1):11-7

Carmichael SW. (2007) Why penguin beaks are sexy! *Microscopy today.* 15 (1): 3

Carroll J, Murphy CJ, Neitz M, Ver Hoeve JN, Neitz J. (2001) Photopigment basis for dichromatic color vision in the horse. *Journal of Vision.* 1(2):2, 80-87

Cloney RA. & Florey E. (1968) Ultrastructure of cephalopod chromatophore organs. *Z. Zellforsch. Mikrosk. Anat.* 89:250–280;

Cronin T.W., Marshall, N.J. (1989) A retina with at least ten spectral types of photoreceptors in a mantis shrimp. *Nature.* 339, 137 - 140.

Cronin TW, Caldwell RL, Marshall J. (2001) Sensory adaptation. Tunable colour vision in a mantis shrimp. *Nature.* 411(6837):547-8;

Cronin TW, Marshall J. (2001) Parallel processing and image analysis in the eyes of mantis shrimps. *Biol. Bull.* 200(2):177-83.

Cuthill IC, Partridge JC, Bennett ATD, Church SC, Hart NS, and Hunt S. (2000) Ultraviolet Vision in Birds. *Advances in the Study of Behavior.* 29: 159-214.

Cuthill, IC. (2007) Animal behaviour: strategic signalling by cephalopods. *Current Biology.* 18, 1059-1060.

Darwin, C. (1871) The Descent of Man and Selection in Relation to Sex. Murray, London.

Deacon SW Serpinskaya AS, Vaughan PS, Lopez Fanarraga M, Vernos I, Vaughan KT, Gelfand VI. (2003) Dynactin is required for bidirectional organelle transport. *J. Cell Biol.* 160:297-301

Demski LS. (1992) Chromatophore systems in teleosts and cephalopods: a levels oriented analysis of convergent systems. *Brain Behav. Evol.* 40:141-56.

Dengler, M, & Nitschke, W. (1993) Color stereopsis: a model for depth reversals based on border contrast. *Perception & Psychophysics.* 53, 150-156.

Desimone, R, and Schein, SJ. (1987) Visual properties of neurons in area V4 of the macaque: sensitivity to stimulus form. *Journal of Neurophysiology.* 57, 835-868

Dresp B, Jouventin P, Langley K. (2005) Ultraviolet reflecting photonic microstructures in the King Penguin beak. *Biol. Lett.* 1(3):310-3.

Dresp B, Langley K. (2006) Fine structural dependence of ultraviolet reflections in the King Penguin beak horn. *Anat. Rec. A Discov. Mol. Cell. Evol. Biol.* 288(3):213-22.

Dyck J. 1971) Structure and Spectral Reflectance of Green and Blue Feathers of the Rose-Faced Lovebird (Agapornis Roseicollis). *Biol. Skrifter.* Vol.18 no.2; 5-65

Dyck J. (1976) Structural Colours. *Proc. Int. Orn. Congr.* 16: 426-437

Fasick JI, Cronin TW, Hunt DM, Robinson PR. (1998) The visual pigments of the bottlenose dolphin (Tursiops truncatus). *Vis. Neurosci.* 15(4):643-51.

Finger E and Burkhardt D. (1994) Biological aspects of bird coloration and avian colour vision including the ultraviolet range. *Vision Res.* 34: 1509-1514

Finger, E., Burkhardt, D. & Dyck, J. (1992) Avian plumage colors: origin of UV reflection in a black parrot. *Naturwissenschaften.* 79, 187-188.

Fox, DL. (1979) *Biochromy: the natural coloration of living things.*

Fratzer C, Dörr S, Neumeyer C. (1994) Wavelength discrimination of the goldfish in the ultraviolet spectral range. *Vision Res.* 34(11):1515-20.

Fritsches KA, Partridge JC, Pettigrew JD, Marshall NJ. *(2000)* Colour vision in billfish. *Philos. Trans. R. Soc. Lond. B. Biol. Sci. 355(1401):1253-6.*

Fujii R. (2000) The regulation of motile activity in fish chromatophores. *Pigment Cell Res.* 13:300-19.

Ghiradella H, Aneshansley D, Eisner T, Silberglied RE, Hinton HE. (1972) Ultraviolet Reflection of a Male Butterfly: Interference Color Caused by Thin-Layer Elaboration of Wing Scales. *Science.* 178 (4066):1214-1217

Ghiradella H. (1994) Structure of butterfly scales: patterning in an insect cuticle. *Microsc. Res. Tech.* 27(5):429-38

Glaw, F, Vences, M. (1994) A Field Guide to Amphibians and Reptiles of Madagascar 2nd edition. Köln: Verlag GBR

Guibal, CRC, & Dresp, B (2005) Interaction of color and geometric cues in depth perception: when does "red" mean "near"? *Psych. Res.* 69, 30-40.

Guillaumée M, Liley M, Pugin R, Stanley RP. (2008) Scattering of light by a single layer of randomly packed dielectric microspheres giving color effects in transmission. *Opt. Express.* 16(3):1440-7

Hastings J.W. (1983) Biological diversity, chemical mechanisms, and the evolutionary origins of bioluminescent systems. *J. Mol. Evol.* 19,309

Hausmann, F., Arnold, K. E., Marshall, N. J. & Owens, I. P. F. (2003) Ultraviolet signals in birds are special. *Proc. R. Soc. Lond.,* B 270, 61-67.

Herring PJ. (2000) Species abundance, sexual encounter and bioluminescent signalling in the deep sea. *Philos. Trans. R. Soc. Lond. B. Biol. Sci.* 355(1401):1273-6.

Herring PJ. (1987) Systematic distribution of bioluminescence in living organisms. *J. Biolumin. Chemilumin.* (3):147-63

Hunt S, Cuthill IC, Bennett AT, Griffiths R (1999) Preferences for ultraviolet partners in the blue tit. *Anim. Behav.* 58: 809-815

Hunt, S., Kilner, R.M., Langmore, N.E., Benett, A.T. (2003) Conspicuous, ultraviolet-rich mouth colours in begging chicks. *Proc. R. Soc. Lond. B. Biol. Sci.* 270 Suppl 1:S25-8.

Hunt S, Cuthill IC, Bennett AT, Church SC, Partridge JC. (2001) Is the ultraviolet waveband a special communication channel in avian mate choice? *J. Exp. Biol.* 204:2499-507.

Ito S & Wakamatsu K. (2003) Quantitative analysis of eumelanin and pheomelanin in humans, mice, and other animals: a comparative review. *Pigment Cell Res.* 16:523-31

Jacobs GH, 1993 The distribution and nature of colour vision among the mammals. *Biol. Rev. Camb. Philos. Soc.* 68(3):413-71.

Jacobs GH, Calderone JB, Fenwick JA, Krogh K, Williams GA. (2003) Visual adaptations in a diurnal rodent, Octodon degus. *J. Comp. Physiol. A Neuroethol. Sens. Neural. Behav. Physiol.* 189(5):347-61

Jacobs GH, Deegan JF 2nd. 1994 Sensitivity to ultraviolet light in the gerbil (Meriones unguiculatus): characteristics and mechanisms. *Vision Res.* 34(11):1433-41

Jacobs GH, Deegan JF 2nd. (1999) Uniformity of colour vision in Old World monkeys. *Proc. Biol. Sci.* 266(1432):2023-8

Jacobs GH, Deegan JF 2nd. (2005) Polymorphic New World monkeys with more than three M/L cone types. *J. Opt. Soc. Am. A Opt. Image Sci. Vis.* 22(10):2072-80

Jacobs GH, Deegan JF 2nd. (2003) Cone pigment variations in four genera of new world monkeys. *Vision Res.* 43(3):227-36

Jacobs GH, Fenwick JA, Williams GA. (2001) Cone-based vision of rats for ultraviolet and visible lights. *J. Exp. Biol.* 204:2439-46

Jacobs GH, Williams GA, Fenwick JA. (2004) Influence of cone pigment coexpression on spectral sensitivity and color vision in the mouse. *Vision Res.* 44(14):1615-22.

Jacobs GH, Williams GA. (2007) Contributions of the mouse UV photopigment to the ERG and to vision. *Doc. Ophthalmol.* 115(3):137-44

Jacobs GH, Williams GA. (2006) L and M cone proportions in polymorphic New World monkeys. *Vis. Neurosci.* 23(3-4):365-70).

Jacobs GH. (2008) Primate color vision: a comparative perspective. *Vis. Neurosci.* 25(5-6):619-33.

Jacobs GH. (1996) Primate photopigments and primate color vision. *Proc. Natl. Acad. Sci. U.S.A.* 93(2):577-81

Johnsen A, Delhey K, Andersson S, Kempenaers B. (2003) Plumage colour in nestling blue tits: sexual dichromatism, condition dependence and genetic effects. *Proc. Biol. Sci.* 270 (1521):1263-70

Jouventin, P., Nolan, P.M., Ornborg, J. and Dobson, F.S. (2005) Ultraviolet beak spots in king and emperor penguins. *The Condor.* 107(1):144-150.

Kashina AS, Semenova IV, Ivanov PA, Potekhina ES, Zaliapin I, Rodionov VI. (2002) Protein kinase A, which regulates intracellular transport, forms complexes with molecular motors on organelles. *Curr. Biol.* 14:1877–81.

Kelber A, Vorobyev M and Osorio D. (2003) Animal colour vision – behavioural tests and physiological concepts. *Biol. Reviews of the Cambridge Philosophical Society*, 78:1:81-118.

Kemp DJ, Rutowski RL. (2007) Condition dependence, quantitative genetics, and the potential signal content of iridescent ultraviolet butterfly coloration. *Evolution.* 61(1):168-83

Kinoshita S, Yoshioka S, Kawagoe K. (2002) Mechanisms of structural colour in the Morpho butterfly: cooperation of regularity and irregularity in an iridescent scale. *Proc. Biol. Sci.* 269(1499):1417-21

Kinoshita S, Yoshioka S. (2005) Structural colors in nature: the role of regularity and irregularity in the structure. *Chemphyschem.* 6(8):1442-59

Korzan WJ, Robison RR, Zhao S, Fernald RD. (2008) Color change as a potential behavioral strategy. *Horm. Behav.* 54(3):463-70.

Krech, D., Crutchfield, R.S., Livson, N., Wilson, W.A. Jr., Parducci, A. (1982) *Elements of psychology.* (4th ed.). New York: Alfred A. Knopf. pp. 108-109

Langridge, KV, Broom, M, and Osorio, D. (2007) Selective signalling by cuttlefish to predators. *Current Biology.* 18, 1044-1045

Land M.F. and Nilsson D.E. (2002) *Animal Eyes.* Oxford University Press

Latanov AV, Leonova AYu, Evtikhin DV, Sokolov EN. (1997) Comparative neurobiology of color vision in humans and animals. *Neurosci. Behav. Physiol.* 27(4):394-404.

Li Y, Lu Z, Yin H, Yu X, Liu X, Zi J. (2005) Structural origin of the brown color of barbules in male peacock tail feathers. *Phys. Rev. E Stat. Nonlin. Soft Matter Phys.* 72(1 Pt 1):010902.

Logan DW Burn SF and Jackson IJ, (2006) Regulation of pigmentation in zebrafish melanophores. *Pigment Cell Res.* 19:206-13

Lythgoe JN and Shand J. (1982) Changes in spectral reflexions from the iridophores of the neon tetra. *J. Physiol.*

Marks WB, Dobelle WH, MacNichol EF Jr. *(1964)* Visual pigments of single primate cones. *Science. 143:1181-3.*

Marshall J, Cronin TW, Kleinlogel S. (2007) Stomatopod eye structure and function: a review. *Arthropod. Struct. Dev.* 36(4):420-48

Massaro, M., Lloyd, SD. & Darby, JY. (2003) Carotenoid-derived ornaments reflect parental quality in male and female yellow-eyed penguins Megadyptes antipodes. *Behav. Ecol. Sociobiol.* 55, 169-175.

Mäthger LM, Land MF, Siebeck UE, Marshall NJ. (2003) Rapid colour changes in multilayer reflecting stripes in the paradise whiptail, Pentapodus paradiseus. *J. Exp. Biol.* 206(Pt 20):3607-13;

Matsumoto J. (1965) Studies on fine structure and cytochemical properties of erythrophores in swordtail, Xiphophorus helleri. *J. Cell Biol.* 27:493–504.

McGraw KJ and Nogare MC. (2004) Carotenoid pigments and the selectivity of psittacofulvin-based coloration systems in parrots. *Comp. Biochem. Physiol. B. Biochem. Mol. Biol.* 138: 229-33

McGraw KJ, Mackillop EA, Dale J, Hauber ME. (2002) Different colors reveal different information: how nutritional stress affects the expression of melanin- and structurally based ornamental plumage. *J. Exp. Biol.* 205(Pt 23):3747-55.

Meier BP, Robinson MD, Clore GL. (2004) Why good guys wear white. *Psychol. Sci.* 15: 82-87.

Meier BP, Robinson MD, Crawford LE and Ahlvers WJ. (2007) When light and dark thoughts become light and dark responses: affect biases brightness judgements. *Emotion.* 7: 366-376.

Michiels NK, Anthes N, Hart NS, Herler J, Meixner AJ, Schleifenbaum F, Schulte G, Siebeck UE, Sprenger D, Wucherer MF. (2008) Red fluorescence in reef fish: a novel signalling mechanism? *BMC Ecol.* 8:16

Modarressie R, Rick IP, Bakker TC. (2006) UV matters in shoaling decisions. *Proc. Biol. Sci.* 273(1588):849-54.

Morrison RL. (1995) A transmission electron microscopic (TEM) method for determining structural colors reflected by lizard iridophores. *Pigment Cell Res.* 8(1):28-36.

Mougeot Mougeot F and Arroyo BE. (2006) Ultraviolet reflectance by the cere of raptors. *Biol. Lett.* 2(2):173-6.

Mougeot, F., Redpath, S.M. and Leckie, F. (2005) Ultra-violet reflectance of male and female red grouse, Lagopus lagopus scoticus: sexual ornaments reflects nematode parasite intensity. *J. Avian Biol.* 36, 203–209,

Newman LA, Robinson PR. (2005) Cone visual pigments of aquatic mammals. Vis Neurosci. 22(6):873-9

Örnborg J, Andersson S, Griffiths R, Sheldon BC. (2002) Seasonal changes in an ultraviolet structural colour signal in blue tits, *Parus caeruleus. Behav. J. Linn. Soc* 76: 237-245

Osorio D and Ham AD. (2002) Spectral reflectance and directional properties of structural coloration in bird plumage. *J. Exp. Biol.* 205(Pt 14):2017-27.

Parker AR. (2000) 515 million years of structural colour. *J.Opt. A: Pure Appl. Opt.* 2: R15-R28

Parker AR and McKenzie DR. (2003) The cause of 50 million-year-old colour. *Proc. Biol. Sci.* 270 Suppl 2:S151-3

Parker AR. (1998) The diversity and implications of animal structural colours. *J. Exp. Biol.* 201(Pt 16):2343-7.

Parker AR. (1995) Discovery of functional iridescence and its coevolution with eyes in the phylogeny of Ostracoda (Crustacea). *Proc. Roy. Soc. Lond. B* 262: 349-355

Parker AR, McPhedran RC, McKenzie DR, Botten LC and Nicorovici NA. (2001) Photonic engineering. Aphrodite's iridescence. *Nature.* 409: 36-7

Pearn SM, Bennett AT and Cuthill IC. (2003*) Proc. Roy. Soc. Lond. B* 270: 859-65

Peters A, Denk AG, Delhey K and Kempenaers B. (2004) Carotenoid-based bill colour as an indicator of immunocompetence and sperm performance in male mallards. *J. Evol. Biol.* 17: 1111-20

Poralla J and Neumeyer C. (*2006)* Generalization and categorization of spectral colors in goldfish. II. Experiments with two and six training wavelengths. *J. Comp. Physiol. A Neuroethol. Sens. Neural Behav. Physiol. 192 (5):469-79*

Porras MG De Loof A, Breuer M and Aréchiga H. (2003) Corazonin promotes tegumentary pigment migration in the crayfish Procambarus clarkii. *Peptides.* 24:1581–9;

Price TD. (2006) Phenotypic plasticity, sexual selection and the evolution of colour patterns. *J. Exp. Biol.* 209(Pt 12):2368-76.

Prum R.O., Torres R.H., Williamson S. and Dyck J. (1998) Coherent Light Scattering by Blue Feather Barbs. *Nature.* 396, 28-29;

Prum RO, Quinn T and Torres RH. (2006) Anatomically diverse butterfly scales all produce structural colours by coherent scattering. *J. Exp. Biol.* 209 (Pt 4):748-65

Prum RO and Torres R. (2003) Structural colouration of avian skin: convergent evolution of coherently scattering dermal collagen arrays. *J. Exp. Biol.* 206 (Pt 14):2409-29.

Prum RO and Torres RH. (2004) Structural colouration of mammalian skin: convergent evolution of coherently scattering dermal collagen arrays. *J. Exp. Biol.* 207 (Pt 12):2157-72.

Prum RO, Torres RH, Kovach C, Williamson S, Goodman SM. (1999) Coherent light scattering by nanostructured collagen arrays in the caruncles of the malagasy asities (Eurylaimidae: aves) *J. Exp. Biol.* 202: 3507-3522

Prum RO, Torres RH, Williamson S and Dyck J. (1999) Two-dimensional Fourier analysis of the spongy medullary keratin of structurally coloured feather barbs. *Proc. Roy. Soc. Lond. B* 266: 13-22

Raman CV. (1935) The Origin of the Colours in the Plumage of Birds *Proc. Indian Acad. Sci. Sect.A*, 1-7.

Rick IP, Bakker TC. (2008a) Males do not see only red: UV wavelengths and male territorial aggression in the three-spined stickleback (Gasterosteus aculeatus). *Naturwissenschaften.* 95(7):631-8.

Rick IP, Bakker TC. (2008b) Color signaling in conspicuous red sticklebacks: do ultraviolet signals surpass others? *BMC Evol. Biol.* 8:189.

Roberson D, Davidoff J, Davies IRL, and Shapiro LR. (2005) Color categories: Evidence for the cultural relativity hypothesis. *Cog. Psych.* 50:378-411.

Rodionov VI, Hope AJ, Svitkina TM, Borisy GG. (1998) Functional coordination of microtubule-based and actin-based motility in melanophores. *Curr. Biol.* 8:165-8;

Rowe MP and Jacobs GH. (2004) Cone pigment polymorphism in New World monkeys: are all pigments created equal? *Vis. Neurosci.* 21(3):217-22

Seikai T, Matsumoto J, Shimozaki M, Oikawa A, Akiyama T. (1987) An association of melanophores appearing at metamorphosis as vehicles of asymmetric skin color formation with pigment anomalies developed under hatchery conditions in the Japanese flounder, Paralichthys olivaceus. *Pigment Cell Res.* 1(3):143-51

Senar JC, Figuerola J and Domènech J. (2003) Plumage coloration and nutritional condition in the great tit Parus major: the roles of carotenoids and melanins differ. *Naturwissenschaften.* 90(5):234-7.

Shahidi F, Metusalach and Brown JA. (1998) Carotenoid pigments in seafoods and aquaculture. *Crit. Rev. Food Sci. Nutr.* 38(1):1-67

Shawkey MD and Hill GE. (2005) Carotenoids need structural colours to shine. *Biol. Lett.* Jun 22;1(2):121-4

Siebeck UE, Wallis GM and Litherland L. (2008) Colour vision in coral reef fish. *J. Exp. Biol.* Feb;211(Pt 3):354-60.

Siitari H, Honhavaara J, Huhta E, Viitala J. (2002) Ultraviolet reflection and female mate choice in the pied flycatcher Ficedula hypoleuca. *Anim. Behav.* 63: 97-102.

Sköld HN, Amundsen T, Svensson PA, Mayer I, Bjelvenmark J, Forsgren E. (2008) Hormonal regulation of female nuptial coloration in a fish. *Horm. Behav.* 54(4):549-56.

Snider J Lin F, Zahedi N, Rodionov V, Yu CC, Gross SP. (2004) Intracellular actin-based transport: how far you go depends on how often you switch. *Proc. Natl. Acad. Sc.i U.S.A.* 101:13204-9.

Spaethe J, Tautz J, Chittka L. (2001) Visual constraints in foraging bumblebees: flower size and color affect search time and flight behavior. *Proc. Natl. Acad. Sci. U.S.A.* 98(7):3898-903.

Stevens, M, Cuthill, IC, Alejandro Pàrraga, C, and Troscianko, T. (2006) The effectiveness of disruptive coloration as a concealment strategy. *Prog. Brain Res.* 155, 49-64.

Stuart-Fox, D. and Moussalli, A. (2008) Selection for social signaling drives the evolution of chameleon colour change. *Public Library of Science Biology.* 6, 25.

Sugden D Davidson K, Hough KA, Teh MT. (2004) Melatonin, melatonin receptors and melanophores: a moving story. *Pigment Cell Res.* 17:454-60

Taylor JD. (1969) The effects of intermedin on the ultrastructure of amphibian iridophores. *Gen. Comp. Endocrinol.* 12:405-16.

Valverde P, Healy E, Jackson I, Rees JL, Thody AJ. (1995) Variants of the melanocyte-stimulating hormone receptor gene are associated with red hair and fair skin in humans. *Nat. Genet.* 11:328-30.

Vértesy Z, Bálint Z, Kertész K, Vigneron JP, Lousse V, Biró LP. (2006) Wing scale microstructures and nanostructures in butterflies--natural photonic crystals. *J. Microsc.* 224(Pt 1):108-10.

Vukusic P and Sambles JR. (2003) Photonic structures in biology. *Nature.* 424: 852-855

Vukusic P, Sambles JR, Lawrence CR, Wootton RJ. (2001) Structural colour. Now you see it - now you don't. *Nature.* 410: 36.

Vukusic P, Sambles JR, Lawrence CR, Wootton RJ. (2002) Limited-view iridescence in the butterfly Ancyluris meliboeus. *Proc. Biol. Sci.* 269(1486):7-14.

Vukusic P, Wootton RJ, Sambles JR. (2004) Remarkable iridescence in the hindwings of the damselfly Neurobasis chinensis chinensis (Linnaeus) (Zygoptera: Calopterygidae). *Proc. Biol. Sci.* 271(1539):595-601.

Wakakuwa M, Kurasawa M, Giurfa M and Arikawa K. (2005) Spectral heterogeneity of honeybee ommatidia. *Naturwissenschaften.* 92(10):464-7.

Wakakuwa M, Stavenga DG and Arikawa K. (2007) Spectral organization of ommatidia in flower-visiting insects. *Photochem. Photobiol.* 83(1):27-34.

Ward Ward MN, Churcher AM, Dick KJ, Laver CR, Owens GL, Polack MD, Ward PR, Breden F, Taylor JS. (2008) The molecular basis of color vision in colorful fish: four long wave-sensitive (LWS) opsins in guppies (Poecilia reticulata) are defined by amino acid substitutions at key functional sites. *BMC Evol. Biol.* 8:210.

Wehner R and Bernard GD. (1993) Photoreceptor twist: a solution to the false-color problem. *Proc. Natl. Acad. Sci. U.S.A.* 90(9):4132-5.

Williams GA, Calderone JB and Jacobs GH. (2005) Photoreceptors and photopigments in a subterranean rodent, the pocket gopher (Thomomys bottae). *J. Comp. Physiol. A Neuroethol. Sens. Neural Behav. Physiol.* 191(2):125-34.

Wong TH, Gupta MC, Robins B, Levendusky TL. (2003) Color generation in butterfly wings and fabrication of such structures. *Opt. Lett.* 28(23):2342-4.

Yin H, Shi L, Sha J, Li Y, Qin Y, Dong B, Meyer S, Liu X, Zhao L, Zi J. (2006) Iridescence in the neck feathers of domestic pigeons. *Phys. Rev. E. Stat. Nonlin. Soft Matter Phys.* 74(5 Pt 1):051916.

Zi J, Yu X, Li Y, Hu X, Xu C, Wang X, Liu X, Fu R. (2003) Coloration strategies in peacock feathers. *Proc. Natl. Acad. Sci. U.S.A.* 100(22):12576-8.

INDEX

A

accounting, 26
acid, 10, 58
adaptation, 50
adsorption, 8
adults, 38
advertisements, 41
advertising, 34
aggregation, 49
aggression, vii, 56
algae, 9, 19
alternative, 3, 31
amphibia, 57
amphibians, 10, 11, 12, 19, 24, 35
anatomy, 21
animals, vii, xi, 7, 8, 9, 10, 16, 19, 27, 29, 30, 35, 36, 39, 52, 53
Aristotle, 1, 25
arousal, 37
arthropods, 10
atoms, 8

B

backscattering, 12
bacteria, 9, 19
beetles, 19
behavior, 57
bioluminescence, xi, 18, 19, 36, 37, 39, 52
birds, vii, xi, 5, 7, 9, 10, 11, 13, 15, 19, 22, 24, 31, 34, 52
blind spot, 21, 26
blindness, 22
blood, 9, 11, 44
bonds, 8
brain, xi, 1, 2, 3, 21, 25, 26, 27, 30, 42, 45
brain structure, 27
breeding, 10, 31

C

carbon, 10
carotene, 9
carotenoids, 7, 8, 9, 10, 15, 16, 47, 56
cartilaginous, 19
categorization, 55
cell, 12, 16, 18, 26, 35
cell membranes, 18, 26
changing environment, 34
channels, 30, 33
chemical structures, 7
chromatophore, 8, 50
classes, 8, 10, 24
classical conditioning, 31
coding, 2
cognition, 43
cognitive process, ix
cognitive science, ix
cognitive system, 34

collagen, 13, 56
commerce, 45
communication, 13, 19, 36, 37, 39, 40, 52
complex interactions, 45
components, 4
composition, 8, 33
compound eye, 25, 40
compounds, 9
concrete, 22
confidence, 44
connectivity, 33
control, 8
copper, 10
coral reefs, 31
cornea, 22
cortex, 26
crystal structure, 50
crystalline, 12
crystals, 2, 11, 13, 24, 57
cues, 40, 51
culture, 43
cuticle, 12, 51
cytoplasm, 10

D

danger, 44
death, 44
decisions, 49, 54
defence, 37, 39
deprivation, 33
depth perception, 42, 50, 51
derivatives, 7, 9, 11
destruction, 44
detection, 29
diet, 9, 10
diffraction, xi, 11
discomfort, vii, xii
discrimination, 29, 30, 31, 34, 51
discrimination tasks, 31
dispersion, xi
disposition, 46
distribution, 49, 50, 52
divergence, 34
diversity, 11, 13, 21, 34, 43, 51, 55

dominance, xi
double bonds, 8
duplication, 22, 34

E

earth, 1, 19
egg, 40
elaboration, xi
electromagnetic waves, 2
electron, ix, 11, 17, 18, 54
electron microscopy, ix
electrons, 8
emission, 19, 25, 39
emotions, 44, 45, 46
energy, 4, 7, 8
environment, 1, 21, 24, 26, 33, 34, 35
equilibrium, 44
evolution, 10, 30, 34, 55, 56, 57
exoskeleton, 9
expertise, ix
exposure, 33
external environment, 26
extrapolation, 33

F

fabrication, 58
factor analysis, 30
family, 19, 31, 34
farms, 47
fat, 47
feelings, 44, 45
females, 23, 31, 38
femininity, 44
fertility, 31
fertility rate, 31
films, 12
fish, vii, xi, 5, 9, 10, 11, 19, 22, 24, 30, 31, 33, 35, 36, 37, 39, 47, 49, 51, 54, 57, 58
fitness, 38
flavonoids, 11
flight, 39, 40, 57
fluorescence, 24, 37, 54

food, 10, 24, 30, 31, 34, 39, 40, 43, 47
Fourier analysis, 56
fovea, 26
France, 47
freshwater, 19
fruits, 34
fungi, 7, 9, 19

G

ganglion, 26
gene, 22, 23, 34, 57
generation, 58
genes, 19, 30, 34
genetics, 53
Germany, ix
gold, 22, 45
granules, 10
Greece, 1
groups, 2, 7, 19, 33
guanine, 12, 24

H

habitat, 34, 38
habituation, 32
harm, 44
harmony, 44
health, xi, 17, 44
heat, 8, 22
heterogeneity, 58
honesty, 44
hormone, 37, 57
hue, 1
human animal, 22, 27, 29, 35, 42, 43
hunting, 32
hydrogen, 8
hypothesis, 34, 37, 56

I

illumination, 19
image analysis, 50
immersion, 45, 46

incidence, 16
innocence, 44
insects, vii, xi, 11, 12, 19, 24, 30, 39, 58
instruments, 4
interaction, 19, 45
interactions, vii, xii, 10, 21, 27, 29, 32, 42
interdependence, 34
interference, xi, 11, 13
intermedin, 57
interrelationships, 30
invertebrates, 7, 11, 49
isolation, 34

J

juveniles, 37

K

keratin, 12, 13, 56
keratinocytes, 16

L

land, 19, 23
language, 43
lattices, 12
laws, 11
learning, 30, 32
light scattering, 13, 56
lithography, 11
love, 44
luciferase, 19
luciferin, 19
luminescence, 19

M

males, 31, 34, 39
mammal, 23
manipulation, 11
mapping, 43
matrix, 13

meals, 40
measurement, 3
media, 33
melanin, 8, 10, 13, 39, 54
melanophore, 49
melatonin, 49, 57
melt, 35
membranes, 17
memory, 45
messages, 36
metamorphosis, 56
mice, 22, 23, 52
microspheres, 11, 51
microstructure, 18
microstructures, xi, 8, 13, 16, 17, 37, 51, 57
migration, 8, 55
mixing, 2, 3
model system, 33
models, 29
modernity, 44
molecules, 7, 8, 9, 10
mood, 8
mosaic, 31
motion, 26, 29, 36
movement, 4

N

nanostructures, 13, 57
NCS, 3
nematode, 55
neon, 39, 54
nervous system, 4
neurobiology, ix, 53
neuronal systems, 45
neurons, 21, 51
next generation, 40
nitrogen, 8
nucleus, 26

O

observations, 32, 41, 42
octopus, 10

ommatidium, 25
optic nerve, 26
optical properties, 13
optimism, 44
organ, xi, 1, 4
organelles, 53
organic compounds, 8
organism, 1, 4, 7, 19
orientation, 4, 12
oxygen, 8, 10, 19

P

parameter, 45
parasite, 55
particles, 12
partition, 43
permit, 29
personality, 46
photons, 8, 26
phototaxis, 29
physical mechanisms, 13
physics, 11
physiology, iv, 8, 33
plants, 7, 8, 9, 11, 19, 40
plasma, 16
plasma membrane, 16
plasticity, 33, 55
platelets, 12
pleasure, 45
polarity, 3
pollination, 40
pollinators, 40
polymers, 8
polymorphism, 56
polystyrene, 11
poor, 22
porphyrins, 9
posture, 4
power, 2, 17, 44
pressure, xi, 13, 14, 21, 24
primate, 22, 30, 42, 53, 54
production, 13, 18
propagation, 41
proteins, 19

psychological association, 45
psychology, ix, 53
psychophysics, 29
pyrimidine, 10

Q

quantum theory, 8

R

radiation, 19
rain, 2, 39
rain forest, 39
range, 2, 4, 15, 16, 22, 24, 30, 31, 33, 34, 39, 40, 44, 51
realism, 45
receptors, 2, 22, 23, 57
recognition, 29
reconcile, 25
redistribution, 35, 37
reflection, 11, 44, 51, 57
refractive index, 16
refractive indices, 3
regulation, 36, 49, 51, 57
rejection, 47
relationship, 33, 40
relativity, 56
reproduction, xi
reputation, 22
resolution, 26
restaurants, 40
retina, 2, 21, 22, 24, 26, 30, 33, 50
rewards, 32
rhodopsin, 26
rings, 10
rodents, 22, 23
rods, 13, 23, 24, 26, 50

S

sadness, 44
safety, vii, xii
salmon, 9, 47
saturation, 1, 4
scatter, 12
scattering, 11, 13, 56
school, 25, 39
search, vii, 57
selectivity, 54
sensation, 3, 44
sensations, vii, xii, 2, 47
sensitivity, 24, 26, 31, 33, 51, 52
sensors, 26
separation, 16
shape, 4, 17, 35, 41
shrimp, 21, 22, 50
signalling, 30, 34, 35, 36, 39, 40, 41, 50, 52, 53, 54
signals, 1, 4, 26, 29, 36, 39, 41, 50, 52, 56
skin, xi, 7, 9, 10, 11, 13, 15, 16, 35, 56, 57
snakes, 22
soil, 44
solid state, 8
speciation, 34
species, vii, xi, 4, 8, 10, 11, 12, 13, 19, 21, 22, 23, 24, 25, 29, 30, 31, 32, 34, 35, 36, 37, 38, 39, 41, 42
specific surface, 31
spectrophotometry, 15
spectrum, 2, 3, 8, 15, 16, 22, 24
speculation, 2, 25
sperm, 55
stability, 44
stages, 33
stars, 19
stimulus, 33, 51
strategies, 39, 58
stress, 8, 36, 54
sulphur, 31
supply, 39
surface layer, 7
survival, xi, 21, 34, 39, 41

T

talent, ix
targets, 35
television, 41

TEM, 54
temperature, 8
territory, 39
three-dimensional space, 34
tissue, ix, 15, 16, 37
traffic, 41
training, 32, 55
traits, 10
transducer, 26
transition, 4, 32
transmission, 16, 17, 51, 54
transmission electron microscopy, 16
transport, 50, 53, 57
tribes, 43
tropical rain forests, 35
tungsten, 4
twist, 36, 58
tyrosine, 10

U

ultrastructure, 57
uniform, 4
universe, 45
urine, 34
UV, 11, 13, 15, 16, 22, 23, 24, 25, 32, 37, 40, 51, 53, 54, 56
UV light, 15, 32

V

values, 2, 3, 32
variation, 22, 31
vegetation, 35
vehicles, 56
vertebrates, 12, 21, 22, 24, 49, 50
video games, 45, 46
vision, iv, 3, 4, 21, 22, 23, 24, 25, 26, 29, 30, 31, 33, 34, 41, 49, 50, 51, 52, 53, 57, 58
visual environment, 27, 33, 43
visual field, xi, 42
visual system, ix, 1, 4, 15, 21, 23, 24, 25, 26, 29, 30, 31, 32, 42

W

wavelengths, xi, 1, 2, 4, 12, 23, 24, 26, 32, 33, 49, 55, 56
wear, 47, 54
wildlife, 24
wind, 1
workers, 12, 29
worms, 19, 39